T0200399

GAMES

The essays from prominent public intellectuals collected in this volume reflect an array of perspectives on the spectrum of conflict, competition, and cooperation, as well as a wealth of expertise on how games manifest in the world, how they operate, and how social animals behave inside them. They include previously unpublished material by former Cabinet minister Sayeeda Warsi, the philosopher A. C. Grayling, legal scholar Nicola Padfield, cycling coach David Brailsford, former military intelligence officer Frank Ledwidge, neuro-psychologist Barbara J. Sahakian, zoological ecologist Nicholas B. Davies, and the final work of the late Nobel laureate Thomas C. Schelling. This is a must-read for anyone interested in the history, nature, and dynamics of games.

DAVID BLAGDEN is Lecturer in International Security and Strategy at the University of Exeter and was previously the Adrian Research Fellow in International Politics at Darwin College, Cambridge. He is widely published in the scholarly and popular press, has served as a witness for several Parliamentary inquiries, and consults for numerous policy organisations. Dr Blagden has also won the Royal United Service Institute's Trench Gascoigne Prize for original writing on defence and security.

MARK DE ROND is Professor of Organisational Ethnography at Judge Business School, University of Cambridge, and a Fellow of Darwin College, Cambridge. A recurring feature in his work is the experience of being human in high-performing environments. His research has been widely reported in the press. His most recent fieldwork involved a world-first attempt to scull the navigable length of the River Amazon unsupported.

THE DARWIN COLLEGE LECTURES

These essays are developed from the 2016 Darwin College Lecture Series. Now in their thirty-third year, these popular Cambridge talks take a single theme each year. Internationally distinguished scholars, skilled as popularizers, address the theme from the point of view of eight different arts and sciences disciplines.
  Subjects covered in the series include

Games

Conflict, Competition, and Cooperation

Edited by

David Blagden

University of Exeter

Mark de Rond

University of Cambridge

with

Janet Gibson

CAMBRIDGE
UNIVERSITY PRESS

CAMBRIDGE
UNIVERSITY PRESS

University Printing House, Cambridge CB2 8BS, United Kingdom

One Liberty Plaza, 20th Floor, New York, NY 10006, USA

477 Williamstown Road, Port Melbourne, VIC 3207, Australia

314–321, 3rd Floor, Plot 3, Splendor Forum, Jasola District Centre, New Delhi – 110025, India

79 Anson Road, #06-04/06, Singapore 079906

Cambridge University Press is part of the University of Cambridge.

It furthers the University's mission by disseminating knowledge in the pursuit of education, learning, and research at the highest international levels of excellence.

www.cambridge.org
Information on this title: www.cambridge.org/9781108447324
DOI: 10.1017/9781108565738

© Darwin College 2019

First published 2019

Printed and bound in Great Britain by Clays Ltd, Elcograf S.p.A

A catalogue record for this publication is available from the British Library.

ISBN 978-1-108-44732-4 Paperback

In memory of

Thomas C. Schelling,

Game Theorist of Conflict
(1921–2016)

Contents

ix

Contents

Figures

Notes on Contributors

David Blagden is University Lecturer in International Security and Strategy at the University of Exeter, affiliated to the Strategy and Security Institute within the Department of Politics. He was previously the Adrian Research Fellow in International Politics at Darwin College, University of Cambridge, where he remains an Associate Member. He obtained his DPhil at the University of Oxford, and has published in journals such as *International Affairs*, *International Security*, and *International Studies Review*.

Mark de Rond is Professor of Organisational Ethnography at the University of Cambridge's Judge Business School and a Fellow of Darwin College. He is the author/editor of eight previous books and many journal articles on topics spanning the fields of organisational science, corporate strategy, management, and leadership. His ethnographic studies include long stints with British military medics in Afghanistan, elite rowers in Cambridge, biochemists in Oxford, and comedians in London; he has also rowed the navigable part of the Amazon. Mark gained his DPhil from the University of Oxford, and consults widely for a variety of private firms, public bodies, and media outlets.

* * *

David Brailsford is Team Principal at Team Sky, the first professional cycling team to support a British rider to victory in the Tour de France. Since that win in 2012, Team Sky competitors have subsequently won the event five more times, making them the most successful Tour team since their formation under his leadership in 2010. He was previously Performance Manager for British Cycling, leading the team to top the cycling medal table at the Olympic Games both in 2008 and 2012, before leaving in 2014 to focus on Team Sky. Brailsford competed as a cyclist himself for four years in

France in his youth, before taking a degree in sports science and psychology at the University of Liverpool. He was made an MBE in 2005, CBE in 2009, and knighted for services to cycling and the London 2012 Olympics in 2013.

Nicholas B. Davies is Professor of Behavioural Ecology in the Department of Zoology at the University of Cambridge and a Fellow of Pembroke College. He is also a Fellow of the Royal Society, and gave the institution's Croonian Lecture in 2015. He is the author of five books; his most recent, *Cuckoo: Cheating by Nature* (Bloomsbury, 2015), won the British Trust for Ornithology's annual best book prize. A past president of the International Society for Behavioural Ecology, he obtained his DPhil from the University of Oxford. He has won awards from the Zoological Society of London, the Cambridge Philosophical Society, the American Ornithologists' Union, and the Association for the Study of Animal Behaviour.

Laure-Sophie Camilla d'Angelo is an Analyst at RAND Europe, based in Cambridge, UK. She holds a PhD in Experimental Psychology from the University of Cambridge and a BSc in Pharmacology from University College London. She has published in journals such as *Behavioural Brain Research, CNS Spectrums*, and *The British Journal of Pharmacology*.

A. C. Grayling is Master of the New College of the Humanities and a Supernumerary Fellow of St Anne's College, University of Oxford. He completed his DPhil at the University of Oxford, and then was on the faculty of Birkbeck College, University of London, from 1991 to 2011. He is the author/editor of more than thirty books on various areas of philosophy, focusing on ethics, language, logic, and the history of ideas. A Fellow of the Royal Societies of Literature and Arts, Grayling is also a representative to the UN Human Rights Council, a past chairman of the Man Booker Prize judging panel, a former Honorary Secretary of the Aristotelian Society, and a Vice-President of the British Humanist Association.

Frank Ledwidge read law at the University of Oxford, then practised as a barrister in Liverpool. He subsequently served as a military intelligence officer on British operations in Bosnia, Kosovo, and Iraq, before working as a civilian justice adviser for UK Missions in Afghanistan and Libya. He holds a PhD from King's College London and now teaches for the University of Portsmouth at the RAF College. He is the author of *Losing Small Wars: British Military Failure in Iraq and Afghanistan* (Yale, 2011), *Investment in Blood: The True Cost of Britain's Afghan War* (Yale, 2013), *Rebel Law: Insurgents, Courts* *and Justice in Modern Conflict* (Hurst, 2017), and *Aerial Warfare: The Battle for the Skies* (Oxford, 2018).

Nicola Padfield is Master of Fitzwilliam College, Cambridge, and a Reader in Criminal and Penal Justice at the Law Faculty, University of Cambridge, where she has worked for more than 20 years. The author/editor of nine books, she has a broad research lens, engaged both in 'hard' law and in socio-legal–criminological research. Her particular recent focus has been on sentencing law and prison release/recall, an area in which she has contributed to many official reports and inquiries. A barrister by training, she has published widely on many aspects of criminal law, sentencing, and criminal justice. She sat as a Recorder (part-time judge) in the Crown Court from 2002 to 2014 and is a Bencher of the Middle Temple.

Barbara J. Sahakian is Professor of Clinical Neuropsychology at the University of Cambridge Department of Psychiatry and Behavioural and Clinical Neuroscience Institute. She is also an Honorary Clinical Psychologist at Addenbrooke's Hospital, Cambridge, and a Fellow of Clare Hall. She is a Fellow of the British Academy and the Academy of Medical Sciences. A past president of the International Neuroethics Society and the British Association of Psychopharmacology, she holds PhD and DSc degrees from Cambridge. She is also a member of the World Economic Forum on the Future of Neurotechnologies and Brain Science and of the Clinical Advisory Board of the Human Brain Project. She is Senior Consultant for Cambridge Cognition, and co-inventor of the widely used CANTAB cognitive function tests. She is also co-inventor of the University of Cambridge/PEAK Advanced Training Programme and Wizard Apprentice.

George Savulich is a Research Associate in the Department of Psychiatry and MRC/Wellcome Behavioural and Clinical Neuroscience Institute at the University of Cambridge, working with Barbara Sahakian. He holds a PhD from the Institute of Psychiatry, Psychology and Neuroscience (King's College London), and has published in journals such as *Translational Psychiatry* and the *International Journal of Neuropsychopharmacology*.

Thomas C. Schelling (1921–2016) was Distinguished Service Professor at the University of Maryland's Department of Economics and School of Public Policy (1990–2005). He obtained his PhD from Harvard University in 1951, and was subsequently on the faculties of Yale (1953–1957) and Harvard (1959–1990), spending the academic year 1958/1959 at the RAND

Corporation. He had previously worked on the staff of the US Marshall Plan (1948–1950) and in the foreign policy team at the White House (1950–1953). Co-awarded the Nobel Memorial Prize for Economic Sciences in 2005, his major works include *The Strategy of Conflict* (Harvard, 1960), *Arms and Influence* (Yale, 1966), and *Micromotives and Macrobehavior* (Norton, 1978).

Sayeeda Warsi is a former UK Cabinet minister. She served as Minister without Portfolio (2010–2012), then as both Senior Minister of State for Foreign and Commonwealth Affairs and Minister of State for Faith and Communities (2012–2014), before resigning from the Government over British policy during the August 2014 Israel–Gaza conflict. She was also Co-Chair of the Conservative Party (2010–2012), becoming the first British-Asian to chair a major UK political party. A former solicitor with the Crown Prosecution Service, she was appointed a member of the House of Lords (as Baroness Warsi) in 2007 and a member of the Privy Council in 2010. She is now Pro Vice-Chancellor of the University of Bolton.

Acknowledgements

This volume – and the 2016 lecture series organised by the University of Cambridge's Darwin College that it is based on – owe a great many debts to a great many people. First and foremost, Janet Gibson (College Registrar at Darwin) is the person who makes everything work: from event logistics to document management, from flagging deadlines to chasing drafts, none of it would have happened without Janet. We are therefore immensely grateful to her, as are all at Darwin whose academic endeavours she has helped over the years – which basically means every fellow and student of the College since she started in 2007.

Espen Koht also deserves a special mention. As Darwin's IT Manager, Espen was instrumental in ensuring that the series was advertised via every available channel, that speakers' audio-visual requirements were met, that the lectures themselves were filmed to a high standard and subsequently made available for download, and that countless other behind-the-scenes tech-support functions were discharged to an exemplary standard.

The senior leadership of the College were invaluable supporters of the series and, therefore, this volume too. The current Master, Mary Fowler, backed the project to the hilt, as well as hosting and introducing our distinguished contributors at a level befitting their professional standing. And Mary's predecessor, Willy Brown, also played an active supporting role: not only did he bring both of us to Darwin as fellows during his tenure, a singular honour in itself, but also he continued to help in numerous seen and unseen ways, including via the hosting/introducing of speakers when the current Master was travelling. Andy Fabian, meanwhile – as Chair of the College's Education and Research Committee and

xvii

inaugurator of the first annual thematic lecture series at Darwin back in 1986 – was a backer and supportive presence throughout.

Numerous other staff and students of the College and University helped with the organisation of the series in an array of capacities – lecture ushers, student hosts, camera crew, facilities support, and more besides. Our sincere thanks therefore go to Stephanie Ashenden, Sue Beckwith (and team), Cavan Bennett, Katia Bowers, Jamie Brittain, Michelle Cain, Darragh Coffey, Tony Cox, Erin Cullen, Amélie Deblauwe, Joe Delaney, John Dix, Matthew Edwards, Chiara Giorio, Angela Goncalves, Michael Gormer, Stefan Graf, Alex Grzankowski, Meredith Hadfield, Hatem Hatem, Ivan Higney (and team), Katja Hofmann, Andy Howells, Angela Ibler, Dan Jones, Markus Kalberer, Marion Kieffer, Nicolas Köhler, Torsten Krude, Noelle L'Hommedieu, Tiancheng Li, Inès Lion, Paula MacGregor, Emile Marin, Hannah McCarthy, Ian McConnell, Tim Milner, Russell Norman, Michael O'Neill, Stephen Owen, Nitzan Peri-Rotem, Jamie Pilmer, James Poskett, Nebojša Radić, Sam Roberts, Florian Roessler, Derek Scott (and team), Utkarsh Sharma, K. C. Sivaramakrishnan, Shahin Tavakoli, Matt Turpie (and team), Jenneke van der Wal, Jackie Walpole, Ben Watkinson-Powell, Julius Weitzdörfer, Chester White, Roger Whitehead, Joseph Wu, and Amani Zalzali. Our sincere apologies to anyone who has escaped our list (or whose name we have misspelt!). We also thank Steven Holt for his thorough copyediting, Christopher Davis for his swift indexing, and Geetha Williams' team at MPS for their attentive production.

The team at Cambridge University Press have been a supportive, constructive, and – most crucially – patient source of assistance throughout the editing process. We therefore thank Clare Dennison, Thomas Harris, Esther Miguéliz Obanos, Lucy Rhymer, and all CUP colleagues for their forbearance: we are glad to be able to reward it by eventually delivering this assemblage of fascinating essays by eminent authors.

Both editors thank their families for all the support that brought them to the point of being in the privileged position of convening a distinguished guest lecture series at the University of Cambridge. David Blagden also particularly thanks Helena Mills for editorial assistance and critical feedback on several chapters, as well as Michael Clark, Samantha Nicholls, and Patrick Porter for facilitating – in various ways – a weekly commute between Exeter and Cambridge during the time of the series itself.

Finally, we thank the chapter contributors themselves. All are distinguished professionals with unique expertise and demanding schedules, so we are grateful to them for investing time and effort both in the original lectures and in the subsequent book chapters. One of them is singled out for particular recognition, however, as you will see from this book's dedication and introduction.

Introduction

Games: The Spectrum of Conflict, Competition, and Cooperation

DAVID BLAGDEN AND MARK DE ROND

Games are ubiquitous. They shape our leisure and our work, our governments and our global relations. They shape our families, our love-lives, and, ultimately, our reproduction – just as they do for all social species. Games consume resources, but also generate them; they foster friendships, deepen enmities, and shed light on both.

For many Britons, the word 'games' invokes a particular sort of early-life memory: triumph or tears on some windswept rugby or hockey pitch, perhaps, during the eponymous school sports lesson of that name. Yet they are not mere child's play. The organisers of the 2014 football World Cup estimate that the tournament reached 3.2 billion viewers – some 44% of the world's population at the time – with more than 1 billion tuning in for the final match alone.[1] Almost 14% of the humans then on Earth, in other words, devoted time and more-than-a-little emotion to watching eleven Germans and eleven Argentines chase an inflatable bag between two nets, following rules first codified on Cambridge's very own Parker's Piece.[2] The global market for video games was estimated at £91.5bn in 2015, meanwhile, with the very frontier of mankind's technological capacity – an area in which Cambridge, both city and university, has long been at the forefront – driven outwards by the pursuit of

---

[1] FIFA Press Office, '2014 FIFA World Cup™ reached 3.2 billion viewers, 1 billion watched final', 16 December 2015, www.fifa.com/worldcup/news/y=2015/m=12/news=2014-fifa-world-cuptm-reached-3-2-billion-viewers-one-billion-watched--2745519.html [accessed 27 January 2018].

[2] BBC, 'Cambridge ... the birthplace of football?!', 22 September 2009, www.bbc.co.uk/cambridgeshire/content/articles/2006/06/09/cambridge_football_rules_parkers_piece_feature.shtml [accessed 27 January 2018].

ever-more-stimulating computer systems.³ And states vie with each other to host the modern Olympic Games, parting willingly with eye-watering sums for the prestige and esteem associated with such a global spectacle. London 2012's £9.3bn price-tag – hardly inconsequential in a country then dealing with significant fiscal overstretch – pales in comparison with the >$40bn and >$50bn that China and Russia, both less developed countries eager for international status, splashed out on Beijing 2008 and Sochi 2014 respectively.⁴

The etymology of 'game' in contemporary usage originates from *gamen* – Old English for 'joy, fun, amusement' – a term itself derived from Norse and Saxon forebears.⁵ Yet many games are not fun at all: they are played in deadly earnest, and for high stakes. Even the original Olympiads, the progenitor of today's organised interstate sporting contests, were valuable means for rival Greek city-states to assuage political pressure for competition and supremacy while preserving mutually useful military and economic cooperation.⁶ Viewers of the cult US television drama *The Wire*, widely hailed for its Dickensian depiction of crime and poverty in post-industrial America,⁷ will be familiar with the refrain 'it's all in the game'. This wry quip, on the part of drug-dealers and the police officers who chase them, reflects a shared understanding

---

³ *Newzoo.com*, 'Newzoo's top 100 countries by 2015 game revenues', 15 October 2015, https://newzoo.com/insights/articles/newzoos-top-100-countries-by-2015-game-revenues/ [accessed 27 January 2018].
⁴ BBC, 'London 2012: Olympics and Paralympics £528mn under budget', 19 July 2013, www.bbc.co.uk/sport/olympics/20041426 [accessed 27 January 2018]; Charles Riley, 'Beijing had few rivals for 2022 Olympics due to cost', *CNN Money*, 31 July 2015, http://money.cnn.com/2015/07/31/news/winter-olympics-2022-beijing/index.html [accessed 27 January 2018].
⁵ *Online Etymology Dictionary*, www.etymonline.com/word/game [accessed 27 January 2018].
⁶ Kostas J. Gallis, 'The Games in Ancient Larisa: An Example of Provincial Olympic Games', in Wendy J. Raschke (ed.), *The Archaeology of the Olympics: The Olympics and Other Festivals in Antiquity* (Madison, WI: University of Wisconsin Press, 1988), p. 217; Mogens Herman Hansen, *Polis: An Introduction to the Ancient Greek City-State* (Oxford: Oxford University Press, 2006), pp. 9–10.
⁷ Rob Sheffield, '100 Greatest TV Shows of All Time' (#2. 'The Wire'), *Rolling Stone*, 21 September 2016, www.rollingstone.com/tv/lists/100-greatest-tv-shows-of-all-time-w439520/the-wire-w439640 [accessed 27 January 2018]; *The Telegraph*, 'The Wire: arguably the greatest television programme ever made', 2 April 2009, www.telegraph.co.uk/news/uknews/5095500/The-Wire-arguably-the-greatest-television-programme-ever-made.html [accessed 27 January 2018]; Charlie Brooker, 'Oh, just watch it ...', *The Guardian*, 21 July 2007, www.theguardian.com/media/2007/jul/21/tvandradio.guide [accessed 27 January 2018].

that both are hostage to forces larger than themselves: to the perverse incentives created by capricious institutions, be they narcotics gangs or law-enforcement bureaucracies, which are themselves responding self-interestedly to far-reaching socio-economic and political failure. During the Cold War, the language of 'games' – usually prefixed by that most horror-invoking term of all, 'war' – even entered the lexicon of diplomacy and military strategy. Yet the exercises and simulations that one side might perceive as merely necessary to preserve defence and deterrence, the other side could easily interpret as signals of hostility and aggression – a dynamic that scholars dub the 'security dilemma'[8] – bringing East and West perilously close to catastrophic conflict.[9]

The study of games, then, is the study of social interaction in the face of different incentive structures. Whether it be children chasing a ball around a playground or a firm contemplating how to respond to a newly established rival, two chess grand-masters facing each other across a table or Chinese naval officers modelling what to do in some future confrontation with the USA, all are responding interactively to certain incentives under certain conditions. Sometimes those incentives engender cooperation, as when a detective and an informant – who may otherwise despise each other – come together to convict some individual of mutual concern. Sometimes they produce bounded competition, as when Formula 1 teams pour resources into beating each other while also agreeing on the desirability of petrol-driven motor racing in the face of rival electric-powered series' rise. And sometimes such incentives produce open conflict, as when stags' rutting inflicts grievous injuries or even death on others of the same species – deer that, given local proximity, could well be brothers – for the sake of passing on their genes to a herd of hinds.

The utility and importance of investigating incentive structures gave rise, in turn, to one of the great innovations of twentieth-century social science: the incorporation of mathematically derived 'game theory' into

[8] Robert Jervis, 1978, Cooperation under the security dilemma, *World Politics* 30(2), 167–214.
[9] Arnav Manchanda, 2009, When truth is stranger than fiction: the Able Archer Incident, *Cold War History* 9(1), 111–133. Such concerns have returned today, of course, especially in Eastern Europe and Asia.

the explanation of human behaviour.[10] With its systematic unpacking of decision-makers' options under various conditions – uncertainty over available pay-offs, imperfect information about others' intentions, ambiguity over the number of future interactions, and so forth – game theory has shed remarkable light on the drivers of conflict, competition, and cooperation. We know more today about the causes of war, the behaviour of firms, the bargaining of legislators, and, indeed, any number of daily individual human choices because of game theory's insights. Yet we also now recognise that certain assumptions underpinning many of game theory's most seminal contributions, such as the assumption of human 'rationality' – where rationality is equated with forward-looking profit maximisation – do not fully reflect human behaviour, particularly the many social and cognitive sources of (dis)utility that people 'play games' around.[11] As such, while the term 'game' can be a useful analytical tool and heuristic metaphor, its deployment merits caution.

The chapters in this volume, derived from a lecture series on 'Games' convened by Darwin College at the University of Cambridge during the period January–March 2016, reflect an array of perspectives on the spectrum of conflict, competition, and cooperation – as well as a wealth of expertise on what games look like, how they operate, and how social animals behave inside them. First, former UK Cabinet minister Sayeeda Warsi considers the 'game' of politics – a trait that voters lament, even while placing ever-greater demands on their governmental representatives – and its potential to conflict with the personal principles that draw individuals to the calling of political representation in the first place. She concludes that there comes a point where an individual must withdraw from the political game, even if that means forfeiting the governmental power to advance causes that one values, if one is to retain the personal principles and moral code that led oneself to politics in the first place. Second, Nicola Padfield – a scholar who combines legal and criminological expertise with professional knowledge of Britain's judicial and

---

[10] John von Neumann and Oskar Morgenstern, *Theory of Games and Economic Behavior* (Princeton, NJ: Princeton University Press, 2007 [1944]).
[11] Donald P. Green and Ian Shapiro, *Pathologies of Rational Choice: A Critique of Applications in Political Science* (New Haven, CT: Yale University Press, 1996); Richard H. Thaler, *Misbehaving: The Making of Behavioral Economics* (New York: Norton, 2015).

penal systems – scrutinises the 'game-playing' that afflicts the pursuit of criminal justice. She contends that, despite the well-meaning pursuit of reform, the system still too often resembles a game of 'Snakes and Ladders' both for the victims of crime and for the culprits – achieving justice and rehabilitation requires laborious ascent through the system, while it is all too easy to slide into injustice and relapse. Third, A. C. Grayling – a leading philosopher of language, logic, and the history of both – unpacks the thought of Ludwig Wittgenstein, specifically the multiple explicit and implicit games that the great Cambridge logician was playing. While Wittgenstein is known for his introduction of the 'language-game' concept, whereby language derives meaning not from objective referents but from its usage, Grayling shows that Wittgenstein was also playing with the discipline of philosophy itself, seeking to protect the things he regarded as important – ethics and religion – from the encroachments of reductive scientific attitudes.

Fourth, David Brailsford – the principal coach/manager behind unprecedented Olympic success for British Cycling and subsequent Grand Tour success for Team Sky – turns his eye on the games that must be played within elite sport. He concludes that athletes, helped by their coaches and medics, make their biggest strides in performance through playing around with mindset – marginal gains can be found in fitness, equipment, diet, physiological support, and so forth, but the 'inner chimp' must first be willing to undergo privations for the sake of a belief system. Fifth, Frank Ledwidge – barrister and participant-turned-critic in Britain's recent expeditionary wars in Afghanistan and Iraq – eviscerates the UK's post-9/11 return to what was once dubbed 'the Great Game': external powers' direct military involvement in the pursuit of strategic interests in Central Asia. He finds not only that political decision-makers were at fault, as is now ensconced in the popular memory, but also that the British Army's senior officers failed to understand – and therefore failed to appropriately strategise for – the sorts of campaigns they were entering, attributing this failure to a lack of military intellectual culture that he hopes is now shifting. Sixth, eminent neuropsychologist Barbara J. Sahakian – writing here with her research associates Laure-Sophie Camilla d'Angelo and George Savulich – considers ways in which the brain itself can play games with our mental health, an area where human understanding still lags behind our knowledge of physical health. Yet games may also offer a way forward: just as we can train our bodies via exercise to sustain longevity and wellbeing, so too the gaming innovations that Sahakian discusses hold the promise of maintaining and improving brain function, even into later life. Seventh, distinguished zoological ecologist Nicholas B. Davies surveys some of the games that non-human animals play, both within generations and across evolutionary time, driven – as humans are too, of course – by the hope of reproductive success. He reveals complex mixtures of cooperation and competition, arms races and innovations, survival stratagems and sexual trickery, to paint a fascinating picture of life on Charles Darwin's 'entangled bank': individuals, societies, and species locked in conflict for the privilege of replicating themselves.

In the end, we come to our final contributor. Thomas C. Schelling did not invent game theory, but he applied its insights widely to many of the most pressing political, economic, and social challenges of the post-1945 world. Already aged ninety-four by the time of his lecture, Professor Schelling agreed to provide us with a short reflection on outstanding questions arising from that most famous and invoked of game-theoretical heuristics, 'the Prisoners' Dilemma'. That reflection – presented here as our afterword – is his final published work: he passed away in December 2016, some nine months after his lecture, at the age of ninety-five. The material is therefore used by kind permission of his widow, Alice, who was an equally vibrant, generous, and insightful participant in debates with students and staff alike during their stay in Cambridge – a city that Tom first visited, incredibly, on the European staff of the US Marshall Plan in the immediate aftermath of the Second World War. While Professor Schelling published widely, on issues from the public-health implications of smoking to the bargaining problems associated with countering climate change, it is for his work on identifying the conditions necessary for peace to hold in the face of conflictual incentive structures against the backdrop of Cold War nuclear confrontation that he will be most remembered. He was accordingly awarded the 2005 Nobel Memorial Prize in Economic Sciences (shared with Robert Aumann) for 'having enhanced our understanding of conflict and cooperation through game-theory analysis'.[12] With major-power relations now once again souring and the associated spectre of military escalation towards nuclear use returning, a generation of students and policymakers who had hoped that Schelling's insights in that domain could be put away forever are now poring once again over his works. It is only fitting, therefore, that the last word in our volume should go to Tom. And it is only fitting, similarly, that the volume itself should be dedicated to his memory.

---

[12] *NobelPrize.org*, 'Thomas C. Schelling – Facts', www.nobelprize.org/nobel_prizes/ economic-sciences/laureates/2005/schelling-facts.html [accessed 28 January 2017].

1 Personal Principles in the Political Game

SAYEEDA WARSI

If I had been asked to write about games some 30 years ago, you would have received an essay on moves in Monopoly, the language of Ludo, and the detail of draughts. I could have discussed the need to be nimble-footed in netball and hawkish in hockey, but not in my wildest imagination would the discussion have turned to politics. As the second of five girls born into a traditional Muslim Pakistani immigrant family, games and board games were the safe space that was classified as leisure time, and Games in school was the competitive space in which I never excelled. Games were what we learned about in History when we studied the Greeks, where the only areas we seemed to cover were the Gods and the Olympics. Games, in my coming-of-age year at twenty-one, was that amazing Pakistan versus England cricket world cup final with its nail-biting finish and its heroic-looking players. Fast-forward three decades and the word 'games', for me, conjures up images of dark corners, dodgy deals, and disingenuous dialogue. So where did it all go wrong?

Allow me to start with some definitions, some history, and my own journey into politics. Described altruistically by Aristotle as 'of, for, or relating to citizens', politics was the science to create the wellbeing of citizens.[1] In modern terms, the definition of politics has been more broadly cast to include the activities associated with the governance of a country or other area, the debate within political ideologies, or conflict between parties having or hoping to achieve power.

For me, politics represented the journey from activism to application and from interpreting and applying the law to making it. I spent my

[1] Ernest Barker, *The Political Thought of Plato and Aristotle* (Mineola, NY: Dover, 1959), p. 10.

8

twenties qualifying and practising as a lawyer, specialising in criminal defence, mental health tribunal representation, and human rights law. I learnt the art of making a case to fit the rules, but with the sense that sometimes the rules themselves could be better. I volunteered with the Racial Equality Council and the Joseph Rowntree Trust. I campaigned for greater participation of ethnic minorities in public life. I felt able to reconcile my identity as a British Asian with affection for my Pakistani origin. Although there were still challenges for ethnic minorities in Britain, I felt that we were heading in the right direction. I felt that, although there were laws and rules that could be improved, for the most part living by the rules came easily.

And then the rules of the game changed. Most of us can recall where we were and what we were doing as the horror of 9/11 unfolded. We can recall the imagery of the twin towers falling, but what I also recall is the vivid use of language by politicians. As George W. Bush said, infamous and often quoted though it is: 'Every nation in every region now has a decision to make. Either you are with us, or you are with the terrorists.'[2] This signalled the start of the war on terror, and we still feel the after-shocks of that today.

It was clear that identity was to be redefined. There was less scope for nuance, and what defined 'otherness' was now religion. Having decided that these battles so recently fought seemed too hard to fight again so soon, I left Britain in 2002. I travelled to the villages where my parents had come from in the early 1960s and spent a year there. Among other things, I set up a women's empowerment charity. But, having spent time away, I realised that we all had a responsibility to play a part, however small, in ensuring that the atmosphere of division did not grow. At that time I felt that running a campaign or practising the law was not where real change could be made and that to make long-term change one had to be a part of the place where decisions were made. I therefore threw my hat into the ring to fight for a seat in Parliament, and so the journey into politics began.

---

[2] George W. Bush, Speech to Joint Session of the US Congress, 20 September 2001, http://edition.cnn.com/2001/US/09/20/gen.bush.transcript/ [accessed 7 November 2017].

The Political Journey

The 2005 general election campaign taught me some early and painful lessons about the game of politics. I had understood that the campaign was like sport – a competition – but I also thought, like sport, that the competition was played out in accordance with rules. It is not often that politicians are open about how they got it wrong, but being out of the game allows one to reflect both on one's achievements and on where things could have been handled differently. In preparation for this chapter I have looked back on what I considered to be either my defining personal principles or the moral norms according to which I conducted my professional life, and how I implemented or departed from these in my political life.

The 2005 general election campaign provides me with one such case study. This was my first election campaign. I had never fought a town or council or county election and yet found myself in a potentially winnable seat against an unscrupulous but seasoned opponent, in the glare of the media spotlight and during a politically toxic period. A combination of a lack of experience and a lack of faith in my own judgement in a new forum persuaded me to outsource the campaign message, which resulted in a campaign that was ruthlessly political and not particularly principled. As someone who had been a hands-on lawyer – from the way in which I prepared a defence case to owning and running my own practice – my first mistake was not to own the campaign and determine the local message. A departure from the norm, my core principle, led to a campaign of which I regret parts.

My second mistake was to focus on short-term and personal gains rather than remain true to the definition of politics as of, for, or relating to citizens for the result of citizen wellbeing. The winning became the focus, and so I had started to tread the slippery path of 'the end will justify the means'.

Having, as a lawyer, been rooted in the principle of unbiased independent advice and solutions, it seemed that politics required the opposite. The doorstep conversation always and only presented answers to voters from a right-wing ideological position. This was my third departure from the norms or principles I was committed to in my professional life.

These constitute three examples of behaviours that would be completely alien in my professional life having crept into my political life. I explore these not to make judgements about these practices *per se* in politics, but as very personal examples of how the practice of politics lends itself so easily to an alternative way of operating.

I did not win the election. We did reduce the Labour majority, but probably contributed to the increase of the far-right vote because of the relentless focus on immigration both in the Conservative national campaign and in our local campaign. Looking back from a position of principle, I did not deserve to win.

After the elections, I was asked by Michael Howard to take on a Vice-Chairman's role within the Conservative Party. Now, I think it is right to ask that, having not won an election, why was I offered a job? Why did reward follow when the result was not a win? I am sure that there are a number of ways in which the result could have been interpreted as a success due to the reduction in the Labour majority, the effective communications during the campaign, or even access to a more diverse set of voters than the Conservatives had ever managed to engage in the past. But, this being said, we did not win.

Reward Follows a Win

The rules of a game are that reward follows a win. Let me take the Olympics as an example. In historical Greece, the Olympians could be rewarded symbolically by being crowned, financially through prize funds, and politically by raising the trading and negotiating powers of the communities to which they belonged. However, in politics the rules of the game did not apply in the same logical format. It was this anomaly, this illogical application of the rules, that I, as most politicians, was both the beneficiary of, initially, and the victim of in later years. It was a situation repeated often both in Government and in opposition, where appointments, jobs, or rewards – as I call them – did not necessarily follow success. Nor did failure result in consequences. According to the world of business – as my husband often relates to me – politics operates in a world where sales figures and bonus payments are unconnected and no one questions why. What in business would be seen as lunacy is in politics normal and mainly unquestioned.

Where this plays out most starkly and publicly is in a Cabinet or Shadow Cabinet appointment and subsequent reshuffle. The Labour Party's Shadow Cabinet reshuffle of early 2015 is an interesting case in point. It had no logic to why some people were removed and others were not. It had no logic to why some people were rewarded and others were not. It was illogical why people so obviously wrong for a particular subject matter were given that brief. Much of the political commentary featured phrases like 'messy', 'chaos', and 'shambolic'. Putting aside the media biases and political spin – both subjects I will return to – it was obvious that the normal rules of hiring and firing simply did not apply. What should have been the important and serious moment of setting out the team that, as the Opposition, would hold the Government to account to ensure the Government fulfilled its duty to the citizen, instead was presented – both by the Labour Party and externally – as a game engaged in for amusement.

My own experience of a Cabinet reshuffle was similarly chaotic. After the 2010 general election, I was appointed Conservative Party Chairman and a Cabinet Minister, as Minister without Portfolio. The first major reshuffle took place in September 2012. The night before the formal reshuffle began I was invited to a chat with the Prime Minister and offered an alternative role. That role did not excite me and was an offer that I felt was nakedly tokenistic. I refused the new role and indicated that I was happy to step down from Cabinet. The rest of the story I will leave for another place and time. But what was interesting was the number of rules that would in any non-political environment be considered basic requirements but were completely disregarded in this situation. From the disconnect between success and reward to the lack of reasoning for a change, the inappropriateness and informality of the negotiation, to the ultimate appointment of a successor who had been the topic of many discussions over the preceding years about his alleged undermining of the office of the Chairman – not only had the rules not been applied, the book had been well and truly shredded.

The reverse application of this 'no rules' approach in the work-place is the 'job for life' scenario for MPs in safe seats. An even worse manifestation is the House of Lords, the world's second-largest legislative chamber, where it is quite literally a job for life.

There are some behaviours we all instinctively know are wrong, even when we are doing them. Whether it is centuries of religious education teachings rooted in philosophical thinking or the memory of the clip round the back of the ear from a parent that conditions us, we know that it is wrong. Telling a lie, not telling the truth, or maybe even not telling the whole story and knowing the consequences of misinterpretation. In political circles this has become an art form.

In the UK, the cult of the special advisor – the 'SpAd' (Figure 1.1(a)) – or the spin doctor is one that was brought right into our living rooms in the form of Malcolm Tucker in the BBC drama *The Thick of It*. Tucker is painted as a dysfunctional individual whose job is to ensure the minister's office functions. But the reality at the 2015 general election was that the three leaders of the three main political parties – David Cameron, Ed Miliband, and Nick Clegg (Figure 1.1(b)) – were all ex-SpAds. Spin has, across the world, become a standard feature in a political environment. The public seem simply to accept it and huge careers have been built on it. Yet, when we unpick the values or principles that both underpin it and are displayed in the practice of it, we would instinctively know it is what, as a child, would have led to a clip around the ear.

The father of the modern spin doctor – Edward Bernays – was the son of an Austrian Jew who migrated to the USA. He used what were widely identified as propaganda techniques, though he called them 'public relations' due to the negative connotations associated with the term propaganda, which are as familiar today as they were over a century ago. I will explore briefly a few of these techniques.

First is the selective presentation of facts. This is seen most starkly during the presentation of the Budget, which, among other things, has led to the setting up of the Office for Budget Responsibility. Second is the evasive denial. A good example of this can be found in Tony Blair's comments in 1997 when he was discussing Labour's university and fees policy, and he said that Labour did not have any plans to introduce fees. Third is the disingenuous apology. This is an apology without an acknowledgement of wrongdoing; it is used just to lance the boil. A politician is sorry, not for the act, but for the offence taken. Sometimes this is

(a)

(b)

FIGURE 1.1 The rise of the SpAd – fiction and fact: (a) Reproduced by kind permission of PRIVATE EYE magazine/Richard Jolley; (b) WPA Pool/Pool/Getty Images News/Getty Images.

done by framing the apology as a passive abstraction – such as 'mistakes were made' – which allows a line to be drawn under a matter without any personal responsibility being incurred. Fourth is burying bad news. This often takes the form of politicians announcing unpopular things at a time when it is believed that the media will focus on other news. In some cases, Governments have released potentially controversial reports before Christmas or summer recess to avoid significant news coverage. Sometimes that other news is deliberately supplied by announcing popular items at the same time as unpopular ones.

Much of this conduct is used extensively at election time as political parties lay out their manifestos – the terms of their contract with the voter. Yet, this should be the time when the role of parties should be to engage in a way, both in conduct and in terms of providing information, which allows the public to make an informed choice. Once again we can imagine that, if these techniques were used at the time of contractual negotiations in the commercial world, litigation would follow. It would effectively be a breach of civil law and, in some cases, may fall into the criminal jurisdiction. So, once again, the game of politics is played by rules that in professional and personal conduct would potentially be deemed against the law, certainly morally wrong, and contrary to what most people would cite as their personal principles.

Theories of Games in Politics

I agreed to write this chapter many, many months ago and, like a lot of things I agree to, I returned to it shortly before delivering it and questioned why I agreed. I usually ask the office whose smart idea it had been to accept a commitment, and almost every time it is of my own doing. But imagine the horror of realising, over the Christmas break, that I had agreed to a 50-minute lecture at the University of Cambridge and a chapter for a Cambridge University Press book, both with audiences full of lots of clever people and among a line-up of decorated academics who tackle the same subject in the other chapters of the book. Therefore, allow me to supplement my practitioner's chapter with some theories.

Game theory is the science of logical decision-making in rational beings, with extensive applicability to politics. Researchers, pollsters, and strategists have developed numerous game-theoretical models that use different players from the game of politics to predict and inform areas such as output of policy positions, voting behaviours, and diplomatic outcomes. An early application of game theory was Anthony Downs' *An Economic Theory of Democracy*, which is still relevant today, over half a century on, to understanding political ideologies, two-party democracies and coalitions, and convergence of the centre ground to keep winning elections.[3] It was something Tony Blair almost perfected and has led to the emergence and growth of new political parties such as UKIP and the SNP. It also provides an interesting theory of how a misinformed electorate can provide fertile ground for more entrenched and pointed political and ideological positions.

I want to expand a little on a game theory that, in addition to its relevance to politics, was a familiar scenario in my previous life as a criminal defence lawyer and is an interesting prism within which to explain an aspect of government policy: integration and extremism. This takes me back to my early days in politics and, I believe, explains the potential long-term damaging effects of the wrong policy position.

The Prisoners' Dilemma is a game analysed in game theory that shows why two completely 'rational' individuals may not cooperate, even if it appears that it is in their best interests to do so. It was originally framed in this way by Merrill Flood and Melvin Dresher, working at RAND in 1950.[4] In their framing, regardless of what the other decides, each prisoner gets a higher reward by betraying the other ('defecting').

The reasoning hinges on an argument by dilemma: B will either cooperate or defect. If B cooperates, A should defect because going free is better than serving one year in prison. If B defects, A should also defect, because serving two years is better than serving three. So, either way, A should defect. Parallel reasoning will show that B should also defect.

In traditional game theory, some very restrictive assumptions about prisoner behaviour are made. It is assumed that both understand the nature of the game and that, despite being members of the same gang, they have no loyalty to each other and will have no opportunity for

<hr/>

[3] Anthony Downs, *An Economic Theory of Democracy* (New York: Harper, 1957).
[4] Saul I. Gass and Arjang A. Assad, *An Annotated Timeline of Operations Research: An Informal History* (New York: Springer, 2005), p. 49.

retribution or reward outside of the game. Most importantly, a very narrow interpretation of 'rationality' is applied in defining the decision-making strategies of the prisoners. Given the conditions and pay-offs outlined in Figure 1.2 below, Prisoner A will betray Prisoner B. The game is symmetric, so Prisoner B should act the same way. Since both 'rationally' decide to defect, each receives a lower reward than if both were to stay quiet. Traditional game theory therefore results in both players being worse off than if each had chosen to lessen the sentence of his accomplice at the cost of spending more time in jail himself. Thus, a more altruistic approach would have led to a better outcome for both.

This is something that I came across regularly in multi-defendant cases, for example those involving the possession of Class A drugs or conspiracy. The practical application of this comes into play when multiple defendants can be charged with the same matter, represented by different solicitors, and sometimes interviewed with varying levels of disclosure. This can be an interesting environment in which to give legal advice.

It has also been proposed that game theory explains the stability of any form of political government. Taking the simplest case of a monarchy, for example, the King, being only one person, does not and cannot maintain his authority by personally exercising physical control over all or even any significant number of his subjects. Sovereign control is instead explained by the recognition by each citizen that all other citizens expect each other to view the King (or other established government)

	Prisoner B stays silent (*cooperates*)	Prisoner B betrays (*defects*)
Prisoner A stays silent (*cooperates*)	Each serves 1 year	Prisoner A serves 3 years Prisoner B goes free
Prisoner A betrays (*defects*)	Prisoner A goes free Prisoner B serves 3 years	Each serves 2 years

FIGURE 1.2 The Prisoners' Dilemma.

as the person whose orders will be followed. Coordinating communication among citizens to replace the sovereign is effectively barred since conspiracy to replace the sovereign is generally punishable as a crime. Thus, in a process that can be modelled by variants of the Prisoners' Dilemma, during periods of stability no citizen will find it rational to move to replace the sovereign, even if all the citizens know they would be better off if they were all to act collectively.

However, what is interesting is that, when presented with the dilemma over and over again – the iterated Prisoners' Dilemma, which in politics occurs in almost every area of policy, e.g. resources in welfare and foreign policy positioning – the strategies that fare best are those that are more altruistically based. Robert Axelrod, in his book *The Evolution of Cooperation*, used this to show a possible mechanism for the evolution of altruistic behaviour from mechanisms that are initially purely selfish by a process of natural selection.[5] By analysing the top-scoring strategies, Axelrod found several conditions necessary for a strategy to be successful. The most important condition is that the strategy must be 'nice' – that is, the player will not defect before their opponent does. Almost all of the top-scoring strategies were nice, therefore a purely selfish player will not 'cheat' on their opponent for purely self-interested reasons. However, Axelrod contended that the successful strategy must not be blindly optimistic; a player must sometimes retaliate. An example of a non-retaliating strategy is 'always cooperate', which is a bad choice since 'nasty' strategies will ruthlessly exploit such players. Successful strategies must also be forgiving. Though players will retaliate, they will once again cooperate if the opponent does not continue to defect. This stops long runs of revenge and counter-revenge, maximising the points the strategy scores. Last, successful strategies should be non-envious – that is, they should not strive to score more than the opponent.

So, we could conclude that better outcomes could be achieved for all if a more principled approach were adopted. I've tried to unpick this in an area of policy which is all too often in the news and is certainly a stated priority in successive UK governments: counter-terrorism. I have spoken in the past about a policy of disengagement adopted by successive

[5] Robert Axelrod, *The Evolution of Cooperation* (New York: Basic Books, 1984).

governments towards the British Muslim community. For nearly a decade, first under pre-2010 Labour and then during the post-2010 Conservative–Liberal Coalition, successive governments have adopted a policy of non-engagement with a wide range of Muslim community organisations and activists. Many groups and individuals have, over time, been seen as 'beyond the pale'. Indeed, the Coalition Government even set up a formal process to act as judge, jury, and executioner on whether a group or individual was someone ministers could engage with. Both the setting up of this group, aptly named 'SOGE' (the Senior Officials' Group on Extremism), and its less-than-impressive non-evidence-based submissions divided colleagues around the Cabinet table.

My view has always been clear. While there are many groups government should not fund or take as partners, I do not believe government should disengage from large sections of any community. So, in government I simply decided to continue engaging; an action that received regular criticism both from government and, sometimes, in the media. My decision was completely justified in the aftermath of the brutal and tragic murder of Drummer Lee Rigby when so many whom government had formally disengaged with stepped up and, in their unequivocal and unconditional condemnation of the actions of the extremists, became part of the solution.

The counter-terrorism dilemma can be considered in two ways because the policy runs on two nodes: one, the balance between security and liberty; and, two, the breadth and depth of Muslim engagement. Here I will examine the second.

Muslim engagement policy choices, it seems to me, boil down to the different approaches to minority rights in left-wing and right-wing parties. Left-wingers are, consistently with leftist political thinking, at ease with diversity and more likely to adopt a 'multicultural' approach resting on recognising difference. Right-wingers, consistently with centre-right political theory, are more likely to adopt assimilationist political ideas prioritising majority values and insisting minorities assimilate.

The policy choices for the left are therefore either to stay the course and engage with a wide range of Muslim organisations (even those with which they may culturally or politically disagree), or to adopt a Tory-esque approach to generate political and electoral advantage. Consider, as examples, Tony Blair's 2006 speech entitled 'Duty to Integrate' and the politics of the 'decent left',[6] also known as Blue Labour due to their mimicry of Tory policies. The policy choice for the right is between continuing with an assimilationist discourse ('muscular liberalism') or adopting a wider, deeper engagement with Muslim communities in an integrationist approach that recognises that integration is a two-way street and a process in which both the majority and the minority are transformed.

The cost of poor choices by Labour or Conservatives is alienation and an environment that is much more fertile for radicalisation. The cost of the right choice by a political party is political disadvantage through criticism from their left or right flanks. For example, Labour have been criticised for allowing minorities to engage in 'identity politics' and the Conservatives would be criticised for doing the same, undermining British culture and letting minorities do what they want. However, the 'right' choice is a sound policy outcome as it would provide a deeper dialogue with Muslim communities that is good for Muslims and for society at large. A focus on the altruistic characteristics that I have referred to above, combined with the principled position of acting for the citizen or for the State as opposed to personal political gain, would, over time, provide a safer state and a decrease of the toxic environment in which radicalisation can breed. I have represented this in Figure 1.3.

I will move now onto my final two areas: the players and my dilemma. First, the players: who are they and does the nature of the game determine who can take part in it? Let us return to Aristotle's citizens. Though Aristotle's initially feels like a very principled definition of politics, this is clouded when we realise that his definition of citizen does not include the poor, the slave, the old, women or children, and not even able-bodied young men who did not have the fortune to be born in Greece. Similarly, only free men who spoke Greek were allowed to participate in the ancient Games in classical Greece. Although universal suffrage in the UK means that we no longer fall short in the way Aristotle's definition does, it is

---

[6] Tony Blair, 'The Duty to Integrate: Shared British Values', Speech for the Runnymede Trust, London, 8 December 2006, http://webarchive.national-archives.gov.uk/20080909022722/http://www.number10.gov.uk/Page10563 [accessed 8 November 2017].

Conservative/Labour	Wider Muslim engagement	Narrow Muslim engagement
Integrationist approach	A deeper, wider dialogue with British Muslim communities Tackling radicalisation	Labour political advantage Muslim alienation/ radicalisation
Assimilationist approach	Tory political advantage Muslim alienation/ radicalisation	'Progressive tests' Muslim alienation/ radicalisation

FIGURE 1.3 Political approaches to Muslim community engagement.

still important to remember that we did until less than a century ago, more than 2,300 years after Aristotle wrote.

However, the modern game of politics still finds that its players are mainly male, and there are aspects of political life that are still charged in a way that can be unwelcoming for women. It is hard to imagine in the present day, but there is still a major British workplace – one of the country's best-known brands – where women have no right to take maternity leave. If they do have children, they tend to find that their shift patterns are deeply inflexible, making it all but impossible to maintain a normal family life at the same time as doing a job properly – a difficult thing when there is still a prevailing assumption that they, and not their husbands, should be the ones to shoulder the greater part of the burden at home. Considering this, it is perhaps not surprising that there are still some men who work there who think they can get away with deeply creepy behaviour – behaviour that, almost anywhere else, would get them sacked. In this workplace it's not even clear whom you would report this to. You may ask: if those women are getting such a crappy deal, and if nothing much is changing, can anyone else not do something about it? Why can Parliament not act, for instance? It is a good question, albeit one with a slightly confounding answer. The reason Parliament cannot act is both simple and complicated. Parliament cannot act because it is the very place with the problem.

FIGURE 1.4 The resignation. BBC Motion Gallery/Getty Images.

Turning, then, to my dilemma; one I faced in the summer of 2014. Resignation. Let me start with a quotation from Martin Luther King: 'There comes a time when one must take a position that is neither safe, nor politic, nor popular, but he must take it because conscience tells him it is right.'[7] For me, that time was August 2014.

This photograph (Figure 1.4) was taken the night before I submitted my resignation. Quite aptly I had been asked to take part in the First World War commemorations at Westminster Abbey, where I was asked to put out one of the burning candles in memory of the fallen. Putting the lights out seemed to be a very poignant moment as I left.

Serving in government meant a great deal to me; the chance to better Britons' lives by influencing and making policy was what drew me to politics in the first place. But the Cabinet's unwillingness to censure Israel's disproportionate military campaign against Palestinians in Gaza that summer left my continuation in government – especially as Minister of State at the Foreign and Commonwealth Office, and thus an official representative of British foreign policy – as morally untenable

[7] Martin Luther King, 'A Proper Sense of Priorities', Speech in Washington, D.C., 6 February 1968, www.patheos.com/blogs/paperbacktheology/2016/01/ten-great-martin-luther-king-jr-quotes-on-non-violence.html [accessed 7 November 2017].

against the compass of my own conscience. While hardly comparable to the sacrifices made by those commemorated at the previous evening's Westminster service, or by Dr King back in 1968, I had reached a point where continuing to play the game may have been safe, politic, and popular – but it would not have been right.[8]

Conclusion

The examples I have cited here, I hope, explain – through a practitioner's lens – how the world of politics operates. This allows us to consider first whether politics is a game, and second – if it is a game – whether it is played by a set of rules that we would recognise in similar scenarios outside of politics. Third, it allows us to consider whether politics is simply about winning, or about winning to achieve an altruistic end. Fourth, we can consider whether it is altruism for the greater good and for the citizen and whether that, in turn, should necessitate that the rules governing politics be both clearer and even of higher order than those that we apply in other settings; that they should be about principles and not just law.

My personal conclusion is that politics, which gives the opportunity for the few to govern the many, needs to both follow the laws we take for granted in our personal and professional lives, and constantly be mindful of rules outside the law but that are accepted norms, such as the rules of natural justice and principles. Only then will we have fairer, greater, and better representation in politics; frank and honest discussion about the issues; outcomes that stand the test of time; and true glory for those who engage in the game of politics. To end, I want to share something from a writer and something from an artist. First, from a sermon in Westminster Abbey in 1925 by Frederick Lewis Donaldson when he spoke of the seven social sins – popularised later that year by Mahatma Gandhi – as

> Wealth without work
> Pleasure without conscience
> Knowledge without character

---

[8] For a full explanation of my resignation motivations, see my interview with Cathy Newman of Channel 4 News, 5 August 2014, www.channel4.com/news/why-i-quit-over-gaza-exclusive-video-with-sayeeda-warsi [accessed 8 November 2017].

Commerce without morality
Science without humanity
Worship without sacrifice
And politics without principles.[9]

I end with an artist's depiction of a one-time government minister taking a stand against another one-time government minister. At least the
former feels that she emerged from the political game with her personal
principles more-or-less intact.

FIGURE 1.5 A principled position. *The Times*/News Syndication.

[9] Frederick Lewis Donaldson, 'Seven Social Evils', Sermon in Westminster Abbey,
20 March 1925, www.goodreads.com/quotes/32234-the-seven-social-sins-are-
wealth-without-work-pleasure-without [accessed 7 November 2017].

2 The Game of Crime and Punishment

NICOLA PADFIELD

In many games, you have a list of players, a book of rules, and, in simple games, luck plays a bigger role than strategy. In the best games – I think – luck and strategy compete. To many people, crime and punishment is a game. Detective novels and murder mysteries are popular, and even board games can involve criminal law and lawyers. For example, *Verdict*[1] is a 'two-player game with one prosecutor and one defender. Fifteen cases are available, and the game revolves around movement of pawns on the board.'[2] Then there are popular video games, such as *Sherlock Holmes: Crimes & Punishment.*[3]

The real world is rather different. Baroness Warsi – herself a criminal lawyer by background – argues in Chapter 1 that, in politics, principles are more important than rules. I would suggest that both are vital in criminal justice. But maybe we are talking here about a different sort of game: games as theatre, where you have players as well as pawns, or games as unfair or unscrupulous manipulations – do the police or criminals or defence lawyers 'game' the system in this sense? In this chapter I shall explore three areas: police powers, the trial, and sentencing, to illustrate an argument that the 'games' that do go on can be a dangerous diversion from justice and fair outcomes.

---

[1] Charles S. Roberts, Tom Nissel, and Ray Theime, *Verdict* [board game] (Baltimore, MD: Avalon Hill, 1959).

[2] Details are available from http://spotlightongames.com/list/law.html [accessible 8 November 2017].

[3] See Craig J. Newbery-Jones, 2015, Answering the call of duty: the phenomenology of justice in twenty-first-century video games, *Law and Humanities* 9(1), 78–102.

Police Powers

Let us start with the police. The game is often one-sided – and not always in the same direction. The rules on the collection and presentation of evidence are complex and raise difficult questions. Professor Andrew Ashworth wrote an important article in 1998, entitled 'Should the police be allowed to use deceptive practices?'[4] His argument was that, in principle, the police should recognise a duty not to use deceptive practices in the investigation of crime. He concluded as follows:

> In those limited circumstances in which deceptive practices or electronic surveillance can be justified, the reasons for the general duty and the exceptions should form part of police training and re-training, with a view to implanting them in police culture. For so long as the restrictions are regarded as pointless or irritating handicaps to the pursuit of proper goals, law enforcement officers will be tempted to try to circumvent them or simply to ignore them.[5]

It is difficult to draft the rules in part because the police are up against both very vulnerable suspects and clever and sophisticated criminals. Can we vary the rules according to the opposition: perhaps a form of handicap, as in golf? It is easy for the police to catch 'hopeless' criminals and often virtually impossible to catch the sophisticated and well resourced. Too many sophisticated criminals go untouched by the criminal justice process. Should we move the goal posts according to the skills and might of the opposition?

My first example of broken rules is *Maxwell* [2010] UKSC 48: here are the careful words of Lord Brown in the Supreme Court:

> The unchallenged findings ... are not just disturbing but quite frankly astonishing ... as a result of his cooperation with the police, Chapman [the supergrass and main prosecution witness at Maxwell's trial] and other members of his family received a variety of benefits which were not merely undisclosed to the CPS or counsel but were from first to last carefully concealed from them. They were benefits which

[4] Andrew Ashworth, 1998, Should the police be allowed to use deceptive practices?, *Law Quarterly Review* 114, 108–140.
[5] *Ibid.*, 139–140.

both contravened the controls designed to preserve the integrity of Chapman's evidence and were in addition inherently improper. Amongst the more surprising were that whilst in police custody Chapman was at various times permitted to visit a brothel, to engage in sexual relations with a woman police constable, to visit public houses, to consume not merely alcohol but also cannabis and even heroin, to socialise at police officers' homes, to enjoy unsupervised periods of freedom, and indeed, throughout the actual period of the appellant's trial, whilst threatening not to give evidence after all, he was permitted long periods of leisure (hours at a time) in places of his choice, ostensibly as 'exercise', and in addition phone calls and visits from his own solicitor (para 77).

The opposing player in this game, Maxwell, was no saint. He was a professional criminal with a history of violent crime, and he was almost certainly guilty of the murder and the two robberies of which he was convicted. But the public interest in convicting those guilty of murder should not override the public interest in maintaining some level of integrity within the criminal justice system.[6] That the police acted as they did is not just astonishing, as Lord Brown says, but deeply shocking. Ashworth's call for ethical as well as legal training and re-training is essential.

Outrageous policing sometimes comes to light in the context of a criminal trial when judges are asked to exclude unlawfully gathered evidence. This is a powerful tool against abuse. But before the defence can challenge the evidence, they need to know of its existence. The prosecution must be forced to disclose its hand. Eight women have, I believe, now received substantial compensation for the 'abusive, deceitful and manipulative' relationships they had with undercover police officers, some of whom became their lovers and, indeed, the fathers of their children.[7] No wonder that the Home Secretary, in July 2015, asked Lord Justice

---

[6] I do not explore the decision in the Supreme Court, which decided 3–2 that a retrial would be appropriate and not an abuse of process (Maxwell ended up pleading guilty at the retrial); another recent example is *Joof* [2012] EWCA Crim 1475, where again the reader gets the feeling that the police wanted to keep a potential witness sweet at all costs.

[7] Rob Evans, 'Police apologise to women who had relationships with undercover officers', *The Guardian*, 20 November 2015, www.theguardian.com/uk-news/2015/nov/20/met-police-apologise-women-had-relationships-with-undercover-officers [accessed 8 November 2017]; see also House of Commons Home Affairs Committee, *Undercover Policing: Interim Report, Thirteenth Report of*

Pitchford to lead an inquiry into undercover policing[8] – the names of Bob Lambert and Mark Kennedy are now notorious. It is turning into a massive inquiry, with 200 people already given 'core participant' status, many of whom will have publicly funded legal representation.[9]

Just as some police officers can behave appallingly, so can other prosecution witnesses. The line between good and bad behaviour is just as difficult to draw. The 'Fake Sheikh' Mazher Mahmoud, the *News of the World* journalist, was something of a hero in some circles for his investigations of cricket match-fixing and celebrity crimes. His evidence led to the conviction of several people prepared to supply drugs to a fake sheikh (for example, John Alford, the actor, in 1999). Alford, real name Shannon, lost his appeals both in England[10] and at the European Court of Human Rights,[11] basically because it was held that he had voluntarily fallen into the trap and the morality of entrapment was not relevant. Yet, in the trial of Tulisa, the singer, in July 2014, the trial judge dismissed the case against her because of the unreliability of Mahmoud's testimony. Explaining his decision to end the trial, Judge Alastair McCreath said that

> Where there has been some aspect of the investigation or prosecution of a crime which is tainted in some way by serious misconduct *to the point* that the integrity of the court would be compromised by allowing trial to go ahead, in that sense the court would be seen to be sanctioning or colluding in that sort of behaviour, then the court has no alternative but to say 'This case must go no further'.[12]

Session 2012–13 (London: Parliament, 2013), www.publications.parliament.uk/pa/cm201213/cmselect/cmhaff/837/837.pdf [accessed 8 November 2017].

[8] UK Home Office, 'Home Secretary announces terms of reference for undercover policing inquiry', HM Government, 16 July 2015, www.gov.uk/government/news/home-secretary-announces-terms-of-reference-for-undercover-polic-ing-inquiry [accessed 8 November 2017]; the Inquiry's own website is at www.ucpi.org.uk [accessed 8 November 2017].

[9] Perhaps not as massive as Judge Lowell Goddard's historic institutional sex abuse inquiry, already running at seven years and many millions of pounds, but maybe it will be.

[10] [2001] 1 W.L.R. 51; [2001] 1 Cr. App. R. 12; [2000] Crim. L.R. 1001.

[11] [2005] Crim. L.R. 133; ECHR; 06 April 2004.

[12] Roy Greenslade, 'Mazher Mahmood: what the judge said in collapsing the Tulisa trial', *The Guardian*, 21 July 2014, www.theguardian.com/media/green-slade/2014/jul/21/mazher-mahmood-ukcrime [accessed 8 November 2017].

Mahmood and his former driver, Alan Smith, have now been charged with conspiracy to pervert the course of justice, for allegations arising out of this trial. The hero has fallen; will his approach be resurrected by others?

The use of cell informants – prisoners prepared to give evidence against fellow detainees – is another dangerous practice, or game.[13] Rather than choose a historic example of an agreed miscarriage of justice, where a conviction has eventually been quashed, I raise a live case here.[14] Michael Stone was convicted of the infamous murder of Lin Russell and her daughter Megan Kent in 1996. I have no idea whether he is guilty, but he has now served nearly 20 years of his mandatory life sentence, and questions about the reliability of his conviction remain newsworthy. His first conviction in 1998 was quashed; and so he faced a retrial. At both trials the prosecution relied heavily on the evidence of a prisoner informant, Damien Daley. Both convictions were by majority verdicts of 10–2. The second conviction was held to be 'safe' as 'it must have been obvious to the jury that Daley was deeply flawed. He was a hardened criminal, who lied when it suited him and he had, on his own admission, taken every type of drug'.[15] In other words, it was for the jury to evaluate the evidence with their eyes open, not blindfolded. Is it relevant that Daley, the informant, has subsequently been convicted of a separate murder (in 2014) and that Stone's supporters have named another man they believe committed the crime? They have recently failed to persuade the Criminal Cases Review Commission to refer the case back to the Court of Appeal, despite discomforting DNA evidence found on towelling and a bootlace at the scene of the crime. I hold no brief for Michael Stone, but his case illustrates two of my key points so far: first, the pressure on the police to secure convictions, and the real difficulties in devising 'fair play' rules, both before and after conviction; and second, the pressure to 'win' and

---

[13] See Emma Slater, 'The miscarriages of justice that have involved prison informants', *Bureau of Investigative Journalism*, 7 October 2012, www.thebureauinvestigates.com/2012/10/07/a-case-to-answer-the-miscarriages-of-justice-that-have-involved-prison-informants/ [accessed 8 November 2017].

[14] Some examples: *Dudley and Maynard* [2002] EWCA Crim 1942; *O'Brien, Hall and Sherwood* [2000] Crim. L.R. 676–677.

[15] [2005] EWCA Crim 105. Another example is *Grant* [2015] EWCA Crim 1815.

Nicola Padfield

the difficulty of drawing clean lines between acceptable and unacceptable 'gaming'. Does the process need tighter rules, more powerful umpires?

The Trial

Let us now move to the courtroom. It is here that the police despair of the games that lawyers play. Defence counsel are there to test the evidence, but do they overstep the mark? What is the point of the trial? In *Gleeson* (2003), Lord Justice Auld cited his own *Report of the Criminal Courts Review*: 'A criminal trial is not a game under which a guilty defendant should be provided with a sporting chance. It is a *search for truth* in accordance with the twin principles that the prosecution must prove its case and that a defendant is not obliged to inculpate himself, the object being to convict the guilty and acquit the innocent.'[16]

That is easily said; but is our so-called adversarial system the best way to search for truth? The system can feel like a game of blind man's bluff. Who is the blind man? Well, the answer is that several of the players are. Most obviously, the jury: they sit and listen, take notes if they want to, and only at the end of the evidence does the judge explain to them what they should have been listening out for all along. It seems nonsense that the jury have to wait until the end to receive the judge's directions on the law applicable in the case.

It is not only the jury who may be left feeling in the dark. It is also the defendant, the witnesses, and the complainants, or victims. There is a vast divide – a 'them and us' – not between prosecutors and defence, but between legal professionals and other court users. It is the court users, even the police, who have secondary status. In an excellent book published in 2015, Jessica Jacobsen and colleagues described how the Crown Court trial is ritualised, theatrical 'structured mayhem' – organised, yet chaotic.[17] They say that the lawyers and the judge appear to be playing a game in which they 'seek to outdo each other with displays of eloquence, quick wittedness and legal knowledge'.[18] Add to this the 'chummy', often

---

[16] [2003] EWCA Crim 3357.
[17] Jessica Jacobson, Gillian Hunter, and Amy Kirby, *Inside Crown Court: Personal Experiences and Questions of Legitimacy* (Bristol: Policy Press, 2015), see Chapter 5.
[18] *Ibid.*, p. 203.

30

jovial, relationship between lawyers and judge and it is not surprising that lay court users feel alienated.[19] The lawyers are performing for judge and jury (and for each other), thus underlining the marginalised, outsider position of the victim, witnesses, and defendant. There is a great quotation on language in the book that highlights this: 'Well, it's posh innit? ... everyone talks in this funky language.'[20]

Many topical examples illustrate the problems of devising fair rules for trials. Let us take the late Lord Janner as a first example. How can it be that a person, who, medical witnesses for both sides agree, lacks the capacity to brief his lawyers, can be required to attend court? In January 2016, the Law Commission published a 300-page report on proposed reforms, most of which is welcome. In the first paragraph they state that 'The aim of the law in this area is to balance the *rights* of the vulnerable defendant who cannot be fairly tried with the *interests* of those affected by the alleged offence and the need to protect the public.'[21] Is 'balancing' the right term? The *rights* of someone who cannot be fairly tried must surely trump the *interests* of complainants.[22]

Take the courtroom itself. Even the layout is bizarre: defendants play a peripheral role, mere observers, sitting at the back of the court, often in a glass box from which it can be difficult to hear and follow proceedings. Their fate is decided by the lawyers some way in front of them, facing away from them and towards the judge. The extent to which defendants feel excluded – especially from the sentencing process – raises important questions of legitimacy and compliance.[23]

Judges play games too. They talk in code, and use words in ways that normal people do not. Law students have to accustom themselves not only to the language but to the 'game' of distinguishing and applying

[19] *Ibid.*, p. 108.
[20] *Ibid.*, p. 100.
[21] Law Commission, *Unfitness to Plead: Summary* (London: Law Commission, 2016), https://s3-eu-west-2.amazonaws.com/lawcom-prod-storage-11jsxou24uy7q/uploads/2016/01/lc364_unfitness_summary_English.pdf [accessed 1 December 2017], p. 1.
[22] It was the late, great Professor Ronald Dworkin who developed a theory of rights as 'trumps'. See Ronald Dworkin, *Taking Rights Seriously*, 2nd edn (London: Duckworth, 1977).
[23] Both of which are important concepts in criminal justice: see, for example, Adam Crawford and Anthea Hucklesby (eds.), *Legitimacy and Compliance in Criminal Justice* (Abingdon: Routledge, 2013).

precedents and case law, to real or hypothetical situations. I mention here the hearsay rule. The basic rule makes sense: you should not be convicted on the basis of an out-of-court statement introduced to prove the truth of the matter; the witness should come to court and be cross-examined on their statement. But it is not as easy as this. What happens if the witness is too frightened to come to court, or is dead? Trial judges may allow a statement to be read to the jury, but then an appeal court has to decide whether they were right and whether the conviction is 'safe'. Let us look briefly at the cases of *Horncastle* (2009) UKSC 14 and *Al-Khawaja* and *Tahery v. United Kingdom* (2012) 54 EHRR 23. I refer here to quite separate 'stories'. One, a doctor convicted of indecent assault on two patients while they were allegedly under hypnosis, one of whom had died before trial. Two, a man involved in street violence between different gangs of Iranian Kurds convicted of wounding with intent. Three, a man convicted of causing grievous bodily harm to an alcoholic, who died not from the wound but from an alcohol-related illness before the trial. In each trial the judge allowed the statement of a key witness to be read to the jury – the witness was not present and not cross-examined. In each, the Court of Appeal held that the jury's convictions were safe. The cases have zigzagged between the Court of Appeal, the European Court of Human Rights, the Supreme Court, and the Grand Chamber of the European Court of Human Rights – and in Tahery's case back again to the Court of Appeal. His conviction was the one that was eventually quashed, more than eight years after he had been convicted. The conviction was not safe when the frightened Iranian had been allowed not to come to court in person. Lawyers still argue about the outcomes and the reasoning of the judges. Different courts are answering different questions and, of course, different judges have different opinions. There is no one right answer.

It can feel like a game to an academic lawyer, waiting for the next appeal to see how the next case will be decided, playing with words. I will mention, as another example, *Jogee*, another controversial area of law. It raises the question, should someone be convicted of murder (with its mandatory life sentence, even for the accomplice) when he merely foresaw that his friend might deliberately stab someone and kill them? Jogee's appeal was argued in the Supreme Court in October 2015, and we criminal lawyers were agog to learn whether the Court would roll back the current, I say unfair, breadth of the law on joint enterprise liability.[24] When we waited six months for the decision of the Supreme Court in *Horncastle*, many of us, myself included, thought the delay was because of an exciting split or division of opinion among their Lordships. The truth was very different. They came up with a unanimous 'blockbuster' designed to put the European Court of Human Rights in its place; to tell them loud and clearly that English law, in their opinion, has enough protections to allow convictions based on the sole or decisive evidence of absent witnesses. Many appeals could go either way. Does that make it a game of chance? There is a certain lottery – who are the judges and what did they have for breakfast?[25]

Sentencing

I turn now to sentencing. In our system, judges and magistrates individualise sentences by trying to give offenders the 'right' sentence, addressing both the harm they have caused and their culpability or blameworthiness. There are many rules and a statutory body, the Sentencing Council, creates complex, binding guidelines around this.[26] Judges increasingly publish their 'sentencing remarks' in high-profile cases on the internet; the audience for this piece of theatre being the defendant, the victim, and also the press.

Despite a deluge of rules, sentencing is not as principled as it might be. How do we justify the extraordinary discount for guilty pleas – one-third off the standard sentence as long as you plead guilty at the first opportunity? How do we justify the totality principle, which, in effect, gives you a discount for bulk offending? For example, say two years is the standard sentence for a certain sort of burglary. A defendant with six counts of burglary on the indictment would not get 12 years; however bad a burglar you are, you are not as bad as a rapist. So some of the individual

---

[24] They did indeed: see [2016] UKSC 8, [2016] UKPC 7, [2016] 2 WLR 681, where the Supreme Court rewrote the law on joint enterprise.

[25] This is not a fanciful comment: see Shai Danziger, Jonathan Levav, and Liora Avnaim-Pessoa, 2011, Extraneous factors in judicial decisions, *Proceedings of the National Academy of Sciences of the USA* 108(17), 6889–6892.

[26] See www.sentencingcouncil.org.uk [accessed 10 November 2017].

sentences would probably be ordered to be served concurrently rather than consecutively.

In recent years, my research has focused on how sentences are played out in practice. Should judges too take release rules into consideration when they impose a sentence? Take, for example, the extended sentence. If a judge decides that the offender is dangerous, which means that he or she poses a significant risk of causing serious harm, the judge may impose an extended sentence.[27] This is a determinate, fixed-term sentence, but with an extended period of supervision in the community of up to five years for violent offenders and eight years for sexual offenders. There are three problems that arise here. First, Parliament keeps changing the release rules: is it the half-way point or two-thirds? Is release automatic or at the discretion of the Parole Board? Most judges say that they should not have to take into account the release rules when sentencing, but is that fair to prisoners, whom we know compare their sentences? Second, are you measuring dangerousness today or when someone is released? I published a critique of the Court of Appeal's judgement in the appeal of a Cambridge doctor sentenced in 2015 for a catalogue of offences against children: is he likely to be dangerous when he is released sometime between 11 and 16 years after conviction, especially since he faces a life-long Sexual Offences Prevention Order (SOPO) or Civil Preventative Order, plus strict licence conditions?[28] Third, many of these offenders do not get the intended extended period of supervision in the community because, having been released, they are swiftly recalled to prison and serve the extended period of supervision in prison. Release is always con- ditional. All offenders released from prison serve a period of time under supervision. This used to be under the supervision of the Probation Service, but probation services have been privatised and fragmented into twenty-one different Community Rehabilitation Companies (CRCs), owned by a variety of different private companies, working alongside a much smaller National Probation Service (NPS). I have written very

[27] A definition introduced in statutory form in the Criminal Justice Act 2003.
[28] Nicola Padfield, 2015, Bradbury: a tangle of extended sentences?, *Sentencing News* 2, 5–8.

sceptically about this in a recent article, and I am certainly not alone in my criticisms.[29]

The prison population of England and Wales today is about 85,000, of whom about 11,800 are serving a life or indeterminate sentence (none of whom know whether or when they will be released). There were 6,356 prisoners in prison as of June 2015 because they had been recalled after release.[30] How do prisoners negotiate their way through the system? With great difficulty. Of course, it makes a difference what security categorisation is imposed on a prisoner and to which prison a prisoner is allocated, but the prisoner rarely has a say. They are subject to what feel like endless risk assessments and queue up for relevant courses, some of which may not be available in the prison in which they are held, even though they have been told they must do them before they will be considered for release. Tough licence conditions will be imposed upon them when they are released, and many are recalled simply for breaching the general condition, which says

> To be well behaved, not to commit any offence and not to do anything which could undermine the purpose of your supervision, which is to protect the public, prevent you from re-offending and help you to re-settle successfully into the community.[31]

Many prisoners are understandably wary of prison psychologists and probation officers: they know they must give the 'right' answer, rather than honest answers. A prison sentence can feel like a game of 'Snakes and Ladders': you work your way up the ladders and slide unexpectedly down the snakes. Luck certainly plays a role. In their chapters, Sayeeda Warsi and Thomas C. Schelling both discuss an economist's game, the 'Prisoners' Dilemma'. I have never liked it. The average prisoner is not an

---

[29] Nicola Padfield, 2016, The magnitude of the offender rehabilitation and 'Through the Gate' resettlement revolution, *Criminal Law Review* (February), 99–115.

[30] The statistics are from Ministry of Justice, *Offender Management Statistics Bulletin, England and Wales: Quarterly April to June 2015* (London: UK Ministry of Justice, 2015), www.gov.uk/government/uploads/system/uploads/attachment_data/file/471618/offender-management-statistics-quarterly-bulletin-april-june-2015.pdf [accessed 10 November 2017].

[31] For more details, see Nicola Padfield, *Understanding Recall 2011*, Research Paper No. 2/2013 (Cambridge: University of Cambridge Faculty of Law, 2013), http://ssrn.com/abstract=2201039 [accessed 10 November 2017].

economist and is usually not well educated; they may well have a low IQ, learning difficulties, and physical and mental health issues. They are not in a position to make rational decisions, are often ignorant of decisions being made around them, and are hopelessly unsupported. I offer a few quotations from some research we conducted in 2011:[32]

> ... if I have to get something sorted out that's wrong in my paperwork, it's my money on the phone, my stamps on the letters, it's me having to find out ...

> They put you in prison, leave you there, and you have to sort it out yourself. The nightmare is not having contact with the right people at the right time. Probation should come and see you straight away, no ifs and buts, and they shouldn't leave until they are sure they are genuine proven reasons. They should investigate.

> I am angry: it is disgraceful: all this is doing is pushing me back into a life which I was trying to get away from ... It is disgusting, it's horrible, it's unlawful ... You have 28 days to appeal and I've lost 3 weeks already trying to get my solicitor on to my PIN [Personal Identification Number].

> I hate prison but I can't cope outside. I feel I'm lost between two places. I want to succeed but it's overwhelming ... It's like they are leaving me here to rot.

> I am just on hold. I have been on hold for nine months now.

Politicians could reduce the prison population. Instead, they cut back on legal aid and create more crimes. One of the most recent is 'controlling or coercive behaviour in intimate or familial relationships', which came into force on 29 December 2015. Is this a useful addition to criminal law? No – it simply increases the difficulties facing police and prosecutors when deciding what charge or charges to lay. The statutory guidance identifies sixty-four other criminal offences that may apply in domestic abuse cases.[33] Domestic violence is a scourge and its dynamics are difficult to police. The evidence reinforces the need for more training, but

[32] *Ibid.*
[33] See Home Office, *Controlling or Coercive Behaviour in an Intimate or Family Relationship: Statutory Guidance Framework* (London: UK Home Office, 2015), www.gov.uk/government/uploads/system/uploads/attachment_data/file/482528/

does not justify new, complicated criminal offences.[34] We should also note that, while introducing this new crime, the Government is also cutting back on legal aid.[35] Victims of domestic abuse have increasingly to rely on what advice they can get from the hard-pressed Citizens Advice Bureau volunteers.[36] It is not just prisoners who cannot get good legal advice.

Politicians are nervous of challenging the press, of leading public opinion. Baroness Warsi discusses in Chapter 1 the topsy-turvy world of politics, and her version of the Prisoners' Dilemma involves a discussion of assimilation and integration.[37] To develop her theme, the number of Muslim prisoners has more than doubled in the past 12 years; to 12,825 at the end of 2016.[38] These Muslim prisoners are far from a homogeneous group: some were born into Muslim families, while others have converted to Islam. Current sentencing policies do lead to alienation and can lead to radicalisation.

How does the politician measure effectiveness? They rightly worry about cost. It cost just over £36,000 on average to keep someone in prison in 2014/2015.[39] This is marginally less than it used to cost: the

Controlling_or_coercive_behaviour_-_statutory_guidance.pdf [accessed 10 November 2017].

[34] HM Inspector of Constabulary, *Increasingly Everyone's Business: A Progress Report on the Police Response to Domestic Abuse* (London: HM Inspector of Constabulary, 2015), www.justiceinspectorates.gov.uk/hmic/wp-content/uploads/increasingly-everyones-business-domestic-abuse-progress-report.pdf [accessed 10 November 2017].

[35] *R (on the Application of Rights of Women) v. The Lord Chancellor and Secretary of State for Justice* [2015] EWHC 35 (Admin) paints a depressing picture of vulnerable women having to fight (often unsuccessfully) for the right for legal support.

[36] See Citizens Advice Bureau, 'Domestic abuse cases up 24 per cent, reports Citizens Advice', 29 December 2015, www.citizensadvice.org.uk/about-us/how-citizens-advice-works/media/press-releases/domestic-abuse-cases-up-24-per-cent-reports-citizens-advice/ [accessed 10 November 2017]. See also their report on victims' need for support: Imogen Parker, *Victims of Domestic Abuse: Struggling for Support?* (London: Citizens Advice Bureau, 2015), www.citizens-advice.org.uk/global/migrated_documents/corporate/domestic-abuse-victims--struggling-for-support-final.pdf [accessed 10 November 2017].

[37] Another account is to be found in Douglas G. Baird, Robert H. Gertner, and Randal C. Picker, *Game Theory and the Law* (Cambridge, MA: Harvard University Press, 1998)

[38] Grahame Allan and Chris Watson, *UK Prison Population Statistics: Briefing Paper Number SN/SG/04334* (London: House of Commons Library, 2017), p. 14, http://researchbriefings.files.parliament.uk/documents/SN04334/SN04334.pdf [accessed 11 December 2017].

[39] Ministry of Justice, *Costs per Place and Costs per Prisoner* (London: UK Ministry of Justice, 2015), p. T1, www.gov.uk/government/uploads/system/uploads/attachment_data/file/471625/costs-per-place.pdf [accessed 11 December 2017]. But

number of staff in the public Prison Service has fallen by 30% in the last five years. Yet the Chief Inspector of Prisons, in his 2015 Annual Report, pointed out the terrible repercussions of this:

> You were more likely to die in prison than five years ago. More prisoners were murdered, killed themselves, self-harmed and were victims of assaults than five years ago. There were more serious assaults and the number of assaults and serious assaults against staff also rose ... overcrowding was sometimes exacerbated by extremely poor environments and squalid conditions. At Wormwood Scrubs, staff urged me to look at the cells. 'I wouldn't keep a dog in there', one told me ... Improvements in health care were undermined by restrictions to the regime and the unavailability of custody staff to provide supervision ... Our judgement that purposeful activity outcomes were only good or reasonably good in 25% of the adult male prisons we inspected is of profound concern. These are the worst outcomes since we began measuring them in 2005–06. The disappointing findings reflected both the quantity and the quality of activity ... The core day was fatally undermined by staff shortages and this affected outcomes in all areas.[40]

Life in prison is no leisure game. Another fundamental difficulty is that there is no agreement on the primary aim of the criminal justice system, or even of prison. At the moment, in English law, we have a somewhat random collection of often contradictory ambitions. Section 142 of the Criminal Justice Act 2003 provides that any court sentencing an offender must have regard to the following purposes of sentencing:

the costs of re-offending are even greater: the cost of re-offending by recent ex-offenders was estimated by the National Audit Office as costing between £9.5 billion and £13 billion!

[40] Nick Hardwick, *HM Inspector of Prisons in England and Wales: Annual Report 2014–15* (Norwich: HM Stationery Office, 2015), www.justiceinspectorates.gov.uk/hmiprisons/wp-content/uploads/sites/4/2015/07/HMIP-AR_2014-15_TSO_Final1.pdf [accessed 11 December 2017]. In a report published in January 2016, meanwhile, Paul Wilson – the Chief Inspector of Probation – concluded that 'The present rather disjointed provision is a long way from the seamless Through the Gate service so essential to the challenge of reducing high reoffending rates for this group': HM Inspector of Probation, Transforming Rehabilitation: Early Implementation 4 – 'An Independent Inspection of the Arrangements for Offender Supervision' (Manchester: HM Inspectorate of Probation, 2016), p. 4, www.justiceinspectorates.gov.uk/hmiprobation/wp-content/uploads/sites/5/2016/01/TransformingRehabilitation4.pdf [accessed 10 November 2017].

(a) The punishment of offenders,
(b) The reduction of crime (including its reduction by deterrence),
(c) The reform and rehabilitation of offenders,
(d) The protection of the public, and
(e) The making of reparation by offenders to persons affected by their offences.

This is not the place to discuss how useful this is in practice, and whether it has led to greater consistency. I would simply suggest that it might be useful to have one overriding ambition – the reduction of re-offending – and I would argue for a right to rehabilitation, which is central to the work of the prisons. Rule Three of the Prison Rules 1999 provides that 'The purpose of the training and treatment of convicted prisoners shall be to encourage and assist them to lead a good and useful life.' It does not always feel that way.

Conclusions

My gallop through the criminal justice system has, I hope, illustrated how difficult it is to have a process that is 'fair' – and even to establish whom it should be fair to. I mentioned at the beginning of this chapter a board game called *Verdict*: the players are a prosecutor and a defender. In the trial process, the prosecutor and the defendant may indeed appear to be the main actors, but I hope my examples here make it clear that criminal justice is not theatre; nor should it be a game of luck, as it can appear to be.

My discussion of police powers showed three things: the pressure on the police to win, the difficulty of devising rules which prevent inappropriate gaming, and the impossibility of drawing clean lines between acceptable and unacceptable gaming. My discussion of the trial focused more on the trial as theatre: the defendant marooned in the back of the court, far from the chummy world of lawyers and judge. But here too it is difficult to devise fair rules that cover every situation. Finally, sentencing is complex and, from the prisoner's point of view, hard to understand and hard to negotiate. It is a one-sided game compounded by ignorance.

It is particularly difficult to ensure fair play when decision-making is opaque, and the rules too complicated. The justice 'game' can be one-sided and levelling the playing field – improving the quality of advice available to different 'players' – is a basic requirement. Central players have every incentive to game the system, yet 'gaming' might be considered the antithesis of a rule-bound pursuit.

3 Wittgenstein's Games

A. C. GRAYLING

My aim in this chapter is to explain what Wittgenstein meant by invoking the concept of games to explain how language has meaning, and to explain the meaning of that, in turn, for Wittgenstein's broader conception of the philosophical enterprise. But I also want to say something about 'Wittgenstein's Games' in a different sense. This chapter therefore explores why it is that Wittgenstein had an interest in the question of language and the importance of understanding meaning, because there was a second motivation that lay behind his work.

Russell, Language, and Logic

Wittgenstein is very much a philosopher's philosopher, in the sense that in order to understand the enterprise on which he embarked (and I should use the plural and say *enterprises*, because there is a dramatic difference between his first and second attempts to explain how language has meaning, and what the implication of that is for philosophy), one has to put it in context. That context is provided by the work of another Cambridge figure of the early twentieth century, Bertrand Russell.

Russell was ambitious to show that you could place mathematics on a logical foundation – a project known as logicism. His endeavour to this end is embodied in the *Principia Mathematica*,[1] which he wrote with Alfred North Whitehead. The result was a failure insofar as the logicist project is concerned – one of the principal reasons why was demonstrated by

---

[1] Bertrand Russell and Alfred N. Whitehead, *Principia Mathematica* (Cambridge: Cambridge University Press, 1910–1913).

Kurt Gödel in the 1930s[2] – but, in the process of writing the *Principia Mathematica*, Russell developed several important philosophical ideas. This is a wonderful example of what the French poet Paul Valéry once said, which is (to paraphrase) that a difficulty is a light, but an insurmountable difficulty can be the very sun in what it otherwise reveals to us when we attempt to overcome it. This was certainly the case with his logicist endeavour, because in the course of it Russell identified a number of approaches, and advanced a number of ideas, which have proved of the first importance in philosophy.

One such advance – and it had important consequences for Wittgenstein – was that Russell came to the conclusion that, in seeking to explain how we connect what we say to what we are talking about,[3] that is, to effect a connection between language and the world, we must respect some basic common-sense intuitions about what the world is like.

Early in his thinking about this matter, Russell had been persuaded by the view of the Austrian philosopher Alexius Meinong that words have meaning by denoting; that the fundamental semantic relation is denotation.[4] So if I use the word 'lectern', the meaning of that word is the object that it refers to. If I say the word 'watch', using it as a noun, then it has the meaning of the object on my wrist used to tell the time. One consequence of this is that, although there are many words in the language that do not denote physical spatio-temporal objects in the world, nevertheless they must, by virtue of the fact that they are meaningful – for example the words 'unicorn' or 'Harry Potter' – denote something. And if the things denoted are not actually existing entities both in space and in time, they must nevertheless in some sense, as Meinong put it, 'subsist' somewhere and somehow else in reality.[5] That is, they must have some kind of metaphysical being in order to serve as the denotata of the words that we use to refer to them.

Russell accepted this view until he realised – he said he was alerted to its unpalatable consequences by yet another Cambridge luminary,

---

[2] Panu Raatikainen, 2005, On the philosophical relevance of Gödel's incompleteness theorems, *Revue Internationale de Philosophie* 59(4), 513–534.

[3] Bertrand Russell, 1905, On denoting, *Mind* 14(56), 479–493.

[4] Nicholas Griffin and Dale Jacquette (eds.), *Russell vs. Meinong: The Legacy of 'On Denoting'* (London: Passim, 2009).

[5] *Ibid.*

G. E. Moore[6] – that this was a most uncomfortable result, given that merely by talking about your infinite number of siblings you thereby bring them into the semi-existence of subsistence, and this makes for an implausibly crowded universe. W. V. Quine later aptly described such a universe as an 'ontological slum'.[7] Russell said that this realisation offended his vivid sense of reality. His alternative was ingenious. He did not abandon what you would think is the culprit here, namely the idea that the meaning of a word is the object it denotes. Instead he argued that apparently denoting words – nouns and names – are not in fact denoting words, and that there are only two genuinely denoting words in the language: the demonstrative pronouns 'this' and 'that', which are guaranteed to denote on every occasion of their use.[8] The rest of the apparently denoting expressions are actually concealed descriptions. This allowed Russell to preserve an account of meaning based on denotation without the attendant inconvenience of the ontological slum. Thus 'unicorn' is shorthand for the x that has the property of being a unicorn or the x that unicorns.

There was another motivation for Russell, which was his wish to preserve bivalence – the principle that our language is such that there are two and only two mutually exclusive alternatives for what we assert about the world, namely that such assertions are either true or false. It seems banal to say that there are just two truth values, truth and falsity, and that if something is not true it is false and if it is not false then it is true. But in fact this is a very consequential view, for there are other possibilities relating to a proposition not being true. It might fail to be true because it is meaningless, or because there may be more than two truth values – indeed, we now have multivalent logics, some with an infinite number of truth values. This would suggest that the genuine opposition in play here is not between 'true' and 'false' but between 'true' and 'not true', and that makes a large difference.

Russell was, however, anxious to preserve bivalence because he wanted the logic that he invoked to explain mathematics and language to be a

[6] G. E. Moore, 1899, The nature of judgement, *Mind* 8(30), 176–193.
[7] Phillip Bricker, 'Ontological Commitment', in Edward N. Zalta (ed.), *The Stanford Encyclopedia of Philosophy* (Stanford, CA: Stanford University Press, 2016), https://plato.stanford.edu/archives/win2016/entries/ontological-commitment/ [accessed 11 December 2017].
[8] Russell, 'On denoting'.

classical two-valued logic. The reason he wanted this, in turn, was that, in order to show that the many nouns and names in our language, other than the demonstrative pronouns, are concealed descriptions, one needs an analysis of the surface forms of language into what logically underlies them, and the logic required for this is the perspicuous language of the first-order predicate calculus. To illustrate: suppose you hear somebody claiming that the present King of France is wise. Now, unless you think Monsieur Macron is the King of France, you will be inclined to think that the French throne is empty at present, and you will therefore be inclined to think that it is not true that the present King of France is wise – not because there is an unwise present King of France, but because there is no such thing, wise or unwise.

What Russell did was to dig into the underlying structure of sentences like 'the present King of France is wise' to show why it is false for a reason other than the unwisdom of someone who fits that description. He did this by saying that, if you examine the logical structure of the sentence, representing it in purely logical terms and thus exposing the form of the thought being expressed by the sentence, you see that you are saying 'There is something, let us call it x, which has the property of being France's present King; and anything whatever which is a present King of France is identical to x; and x is wise.' Thus the analysis yields three conjoined sentences, respectively asserting the existence of something x, the uniqueness of x, and x's possession of a certain property (viz. wisdom). The logical machinery of the second of these sentences – 'anything whatever which is a present King of France is identical with x' – takes care of the word 'the' in English, because 'the' implies uniqueness. So, in order to capture the idea that there is one and only one thing being talked about, you have to say that if anything has that property then it is the same thing as x. This analysis reveals that you have two ways of showing why the original proposition is false. It is false either if there is an unwise King of France at present, or if there is no King of France at present.

This overcomes the problem. You are all familiar with versions of this problem, perhaps, when you are asked a question and there is an unfulfilled presupposition lying behind the question. The classic example is someone A asking someone B 'Have you stopped beating your wife?' Of course, if B never did such a thing he cannot say 'no' because that means he is still doing it, and he cannot say 'yes' because it means that he once did. So he has to say 'Hold on a second, there is an assumption lying behind what *A* says, which is that I once did it or that I still do it, and that assumption is not true.' Therefore, the question cannot be answered. That offers a different way of understanding the idea that the logic of what is said determines its truth-value, though in this case it does so by introducing a third truth-value, viz. neither-true-nor-false – a so-called truth-value gap. In contrast, Russell's way of dealing with the matter preserves bivalence. Yet another Cambridge luminary, Frank Ramsey, described Russell's theory of descriptions as a paradigm of philosophy.[9] It is indeed a perfect example of a piece of conceptual analysis under certain assumptions.

Wittgenstein's Early Influences

I have set out the foregoing because it is what first influenced Wittgenstein when he came to Cambridge to study under Russell. It sparked the work that resulted in the only philosophical book he published in his lifetime, the *Tractatus Logico-Philosophicus*.[10]

Ludwig Wittgenstein was brought up in a wealthy and influential family in Vienna. His father, Karl Wittgenstein, was an industrialist and patron of the arts and music. Karl had tried to educate his children at home, but this did not prove successful; indeed, it was so unsuccessful that Ludwig himself, when he was finally sent to school, was unable to get into a *Gymnasium* – analogous to a UK grammar school – and had to be sent off to Linz to a *Realschule*, the equivalent of what used to be called in England a 'secondary modern'. There he was an immediate contemporary of Adolf Hitler, although there is no suggestion that they knew one another because, though they were exactly the same age, they were in different classes one year apart. Wittgenstein could not get a university place because he did not have sufficient qualifications, so he was sent

---

9 Bernard Linsky, 'Russell's Theory of Descriptions and the Idea of Logical Construction', in Michael Beaney (ed.), *The Oxford Handbook of the History of Analytic Philosophy* (Oxford: Oxford University Press, 2013), p. 407.
10 Ludwig Wittgenstein, *Tractatus Logico-Philosophicus* (London: Routledge, 2001 [1961 edn, trans. D. F. Pears and B. F. McGuinness]).

to a technical college in Charlottenburg near Berlin, where he became interested in aeronautics. This interest arose because he had originally had a desire to study physics with Ludwig Boltzmann in Vienna, and he had heard Boltzmann say that the new science of aeronautics required people who were both heroes and geniuses. This is because, in those early days of heavier-than-air flight, you needed to be brave enough to risk breaking your neck and clever enough to understand the principles of aerodynamics.

Wittgenstein very much wanted to be a hero and a genius. As a result, he became interested in aeronautics and went to study it at Manchester University. Designing a propeller at Manchester made him interested in the mathematical properties of propellers. This led him to an interest in mathematics itself. Wittgenstein soon thought – rather characteristically of him – that he had solved the problem of the foundations of mathematics, and therefore wrote an essay setting out his views, which he sent to Gottlob Frege at Jena University. Frege invited him to come and talk, and when Wittgenstein did so Frege explained the inadequacies of his essay to him, and told him that he needed to study with somebody who knew something about the philosophy of mathematics; he recommended Russell.

Thus it was that Wittgenstein came to Cambridge in 1912 to spend five terms dazzling Russell with his powerful and eccentric personality. Russell reports that Wittgenstein used to knock on his door at two o'clock in the morning and storm into his rooms; Russell would say to him 'Are you thinking about logic or your sins?', and Wittgenstein would say 'Both'.[11] Wittgenstein impressed not only Russell but also G. E. Moore, who said that he thought 'very well' of Wittgenstein because he was the only person who looked puzzled during Moore's lectures.[12]

Russell explained to Wittgenstein what he had attempted in the *Principia*, and talked about the move he was then making into more central areas of philosophy, particularly epistemology. He was at that time writing a book on the subject, parts of which he abandoned as a result of certain criticisms that Wittgenstein offered. But Russell sparked a desire

---

[11] Bertrand Russell, *The Autobiography of Bertrand Russell* (Abingdon: Routledge, 2014 [1950]), p. 330.
[12] *Ibid.*

in Wittgenstein to provide a solution to the problem of how language has meaning and what the underlying logic of language is.

When war broke out in 1914, Wittgenstein joined the Austrian Army. He was a mechanic for two years behind the Eastern Front. He then did officer training and went to the Italian Front as an artillery observer, where, with the surrendering Austrian forces at the end of the war, he was imprisoned at Monte Cassino. He had in his knapsack the manuscript of the *Tractatus Logico-Philosophicus*, a name suggested by G. E. Moore in imitation of one of Spinoza's works.

The *Tractatus*: Wittgenstein's First Cut on Language and Logic

The *Tractatus* sets out, in a very pared-down, austere way, a view about how language and world connect, and therefore how language has meaning. In rough terms, the proposal is this: given that both the world and language are complex things, and given that complexes have structure, an analysis of their structure would terminate in their most basic elements, which by definition are not further analysable. Then you look to see how the two structures parallel one another, even if you cannot specify what the different levels of structure actually are.

The world, Wittgenstein said, consists of *facts* not *things*: if you give a straightforward inventory of the things in the world, such as my left shoe and my right shoe and this book, you would not be describing how the world is. But if you say 'There is my left shoe and my right shoe, and my left shoe is on my left foot', then you have described a *fact*, a *way things are* in the world. So the world is the totality of facts, of how things are arranged, not merely a list of things. Language consists of what we can say about those facts by the use of propositions. By 'proposition', Wittgenstein meant the content of a thought. A proposition is not to be identified with any sentence that can be used to express it; the sentences 'It's raining', 'Il pleut', and 'Es regnet' all mean the same, in different languages, namely that it is raining. They all express the same proposition that it is raining, so it is the proposition which, in its physical embodiment as an uttered or written sentence in a given language, relates to facts.

Now, the question is, how do they relate? You see how when you look at the deeper levels of structure. Facts are made up of what Wittgenstein called 'states of affairs', and states of affairs are made up of 'objects'. He said (and I paraphrase) 'I do not know what states of affairs are, and I do not know what objects are; I do not know what one would empirically discover when investigating what constitutes these lower levels of structure; but any complex has structure and any complex must eventually end in the simplest level of structure. So let us just denominate them as follows: the world is made up of facts, facts are collections of states of affairs, and states of affairs are collections of objects. Parallel to this is a structure of language. Language consists of propositions, propositions are made up of elementary propositions, and elementary propositions are made up of names. These are not names like Tom, Dick, and Harry; they are whatever the most basic units of language are.'

Wittgenstein's claim is that the structures parallel one another: propositions describe facts – actual and possible facts – while elementary propositions describe states of affairs and names denote objects. It is at this very lowest level of structure that the connection between language and world is effected, because the arrangement of names in an elementary proposition is a picture of an actual or possible arrangement of objects that constitute a state of affairs. Wittgenstein meant this idea of a picturing relation to be taken quite literally, in capturing the sense that different formal structures can be representations – re-presentations – or depictions of one another. He gives the example of the logical relations between a musical score, the movement of the pianist's fingers on the keyboard, the airwaves propagating to the eardrums of listeners, and the music as heard. Each of these different instantiations has an internal connection to the others. From the musical score you can deduce something about what will be heard and *vice versa*. Somebody who has the appropriate skill could listen to a piece of music and transcribe it onto staves, paralleling the pianist's skill in translating marks on staves into movements of her fingers and thence sounds in others' consciousness. So, as it were, the musical score is a picture of the sounds heard, and the sounds heard can be treated as a picture of the musical score.

This, accordingly, is the nature of the relation that matches the arrangement of names and elementary propositions to the arrangement of objects. The names picture the objects, they denote them – so the denotative theory of meaning is preserved, but the explanation of how propositions at the top of the structure can be pictures of actual or possible facts resides in that relationship at the bottommost level of the two structures. This is, in essence, what the *Tractatus* says about how language acquires meaning.

But there is a significant further aspect to the *Tractatus*, which Wittgenstein described by saying that the most important thesis offered by the *Tractatus* lies in that half of the book that could not be written. This rather vatic remark indicates the other game Wittgenstein is playing, to which I will return below.

Wittgenstein – again rather characteristically, you will remember that the essay he sent to Frege claimed to solve the problem of the logical basis of mathematics – now claimed that the *Tractatus* solved all the problems of philosophy.[13] He claimed this because, if propositions are pictures of actual or possible facts, then if they are not pictures of actual or possible facts they are meaningless; they are *nonsense* in the literal sense of the term. But the only possible and actual facts in the world are those that can be described by science. So the only meaningful discourse is science (common sense is included in science because science gives a more detailed and circumstantial account of our common-sense beliefs). It follows that all the things that we wish to say in ethics, in aesthetics, and religion, since the propositions or seeming propositions that we assert in connection with those subject matters are not pictures of actual or possible facts, are in the literal sense nonsensical. This does not mean that they are in any way unimportant, says Wittgenstein; and this is his point about the unwritten half of the *Tractatus*. All the things that are most important cannot be said because the propositions in which you attempt to say them are not pictures of actual or possible facts.

The *Philosophical Investigations*: Language as Useful Game

Thus did Wittgenstein think that he had solved all the problems of philosophy. If you understand how language works, you see that you can say

[13] Wittgenstein, *Tractatus* [1961 edn], see 'Preface'.

things of a scientific and common-sense nature, but that you cannot say anything about ethics and aesthetics or anything really important – you can only live them. That therefore solves the whole problem: people can stop doing philosophy now, can shut up shop and go home. Wittgenstein accordingly abandoned philosophy, and became a school teacher as did Russell and Karl Popper, both of whom, shocked by the First World War and its effect on what had seemed to be advanced civilised countries, came to believe that the only way to prevent another tragedy of that kind was through the education of the young.

Wittgenstein was, however, unsuccessful as a teacher. He taught for about four years at different schools in rural Austria, but eventually had to abandon his school-mastering career because he got into trouble, having struck a child who fainted or fell unconscious as a result. Naturally the parents were angry and sought to institute an enquiry against Wittgenstein. He therefore left school-mastering and became a gardener at a monastery in Vienna. He wanted for a time to be a monk.

Wittgenstein was not engaged at that point with philosophical reflection until he was invited by members of the 'Vienna Circle', which included Moritz Schlick, Rudolf Carnap, and Otto Neurath, who were interested in what he had to say in the *Tractatus* because they thought, mistakenly, that it was similar to their view. Their view is called 'logical positivism', and its principal thesis is that only empirically verifiable sentences have meaning, and that what we say in ethics and religion does not have meaning because one cannot verify the claims made in those discourses. Wittgenstein did not accept the invitation to meet with the Circle, although he did discuss privately with Schlick and one or two others of its members. But gradually he was persuaded by these encounters that he had not solved all the problems of philosophy, for he began to see that there are insurmountable difficulties in the *Tractatus* theory.

In 1929, as a result of this, Wittgenstein returned to Cambridge and submitted the *Tractatus* for the degree of PhD. He worked with Frank Ramsey, then only 25 years of age. Tragically, Ramsey died the following year under an anaesthetic at Guy's Hospital in London. He was a brilliant young man who had already made major contributions to philosophy, economics, and probability theory. Ramsey served as Wittgenstein's supervisor, and Russell and Moore examined the *Tractatus*. Moore was opposed to the PhD degree, which he regarded as an unpleasant alien import from Germany via the USA. His report on the *Tractatus* read as follows: '[This] is a work of genius, but it otherwise satisfies the requirements for a PhD.' Wittgenstein was then helped by Russell to get a five-year research fellowship at Trinity College, Cambridge. He began to write copiously. Over the next two decades, the notes he made provided the seam from which his posthumous books (he died in 1952) were mined and edited for publication by his executors. Some of the interim thinking he did was dictated to students and circulated in samizdat form during the 1930s. The most significant of the posthumous works (Wittgenstein scholars regard all of them as significant) for the wider philosophical community is what we now know as the *Philosophical Investigations*.[14]

It is in the *Philosophical Investigations* that we meet the concept of games. It is central to the second and very different phase of Wittgenstein's thought. His revised thinking was premised on the conviction that he had been badly wrong in the *Tractatus*, and indeed states in the exordium to the *Philosophical Investigations* that you have to read the *Tractatus* alongside it in order to see how wrong those earlier ideas were.

One major continuity between the early and the late philosophy is that Wittgenstein, at the time he was writing the notes that were eventually published as the *Philosophical Investigations*, continued to believe that philosophy is not a real pursuit, that philosophical problems are actually spurious problems, that they arise because we misunderstand the way language works. He had taken this view in the *Tractatus*, and still held that view, but Wittgenstein now had a dramatically different way of explaining why. This is that language does not acquire meaning by standing in a relationship to something other than itself through connections of denotation or reference, but instead that meaning arises from the *uses* that we make of words in the very many different ways that language is employed. He saw now, for example, that the *Tractatus* had given us an extremely impoverished view of language, as if language were only ever used to say things like 'This is a table' or 'This is a watch' and 'There is a glass on the table', 'It's raining outside' – as if language were just

---

[14] Ludwig Wittgenstein, *Philosophical Investigations* (Oxford: Wiley-Blackwell, 2009 [1953] (4th edn, eds. P. M. S. Hacker and Joachim Schulte).

a collection of assertions, a set of assertoric propositions. He now recognised that there are many different things that we do with language:
we ask questions, give commands, make statements, express desires and
wishes, lie, pretend, playact, quote poetry, make demands and promises –
and much besides.

This great variety of types of language-use and the differences
between them are what, according to Wittgenstein, create the philosophical difficulties that arise when we make the mistake of assimilating
uses in one area to uses in another, thus misunderstanding the way an
expression works in a given area of language. He called these different
areas of language 'language games'. He did not seek to imply by the word
'game' anything frivolous or unimportant; he meant it as an allusion to
the way a game is an activity that has rules governing what counts as
the proper moves and aims of the game. Think, for example, of chess
or backgammon; they are self-constituting enterprises, in which what
you do in using a piece in the game is set by conventions. Wittgenstein
described language games as being woven into what he called 'a form of
life'.[15] On this view, the *way* that expressions of a language are employed
is constitutive of their meaning. To put it roughly, therefore, the theory
of meaning in Wittgenstein's later philosophy is that *meaning is use.*

A number of consequences flow from this view. The idea of a denoting
link between language and something other than language has gone. It is
no longer the case that a *relation* to something independent of language
confers meaning on linguistic expressions, still less that something
independent of language determines *the meaning* of those expressions.
Rather, language is self-constituting, language games set the meanings
of expressions used in those games, and if you try to use an expression in
a language game where it does not belong – that is, if you take it out of its
context – you are going to misuse the expression, and thus be in danger
of generating a problem. That is the source, as Wittgenstein saw it, of
philosophical dilemmas. Wittgenstein thought that, if you only understand the normal, straightforward surface use of an expression in its own
appropriate language game, you would never fall prey to that temptation.

[15] Ibid., p. 23.

And this, he thought, solved all the problems of philosophy. So, once again, you can shut up shop and go home.

There are a number of other implications of this view. Wittgenstein talks about language games as having what he described as a 'family resemblance' to one another, in just the same way as members of a family have something recognisably similar to each other in cast of expression, colour of eyes, and shape of face; just so do various language games resemble other language games. But he used the word 'games' for a particular reason, which is that, among all the different games there are, there is no one thing that all games have in common. You will be kept awake forever by the task of trying to find one single thing that all games have in common and which makes them games – an essence or definition of gamehood. The way Wittgenstein put it is that apart from family resemblances between games there is no single criterion, no overarching definition, that you can give of what a game is because of their great variety. This is how it is with language. The different activities that we engage in linguistically have family resemblances to one another, but we cannot assimilate uses in one to another, any more than you can try to apply some of the rules of backgammon to chess. Hence the putative solution Wittgenstein offered to all the problems of philosophy: that they are all just mistaken applications of expressions that do not belong where you think they do.

'Thereof We Must Be Silent': Matters Too Important for Language?

It turns out that at the very end of his life Wittgenstein turned his attention to a traditional philosophical problem which is central to the theory of knowledge, namely, the problem of doubt and certainty and their relation to knowledge claims. In doing this he was addressing some of the work that G. E. Moore had undertaken half a century before. Wittgenstein's notes on this subject were collected and arranged by his editors and published posthumously as a short book entitled *On Certainty*.[16] In it he reprised views that had been advanced in related forms

---

[16] Ludwig Wittgenstein, *On Certainty* (Oxford: Blackwell, 1969 [trans. G. E. M. Anscombe]).

by earlier thinkers whose work he knew only imperfectly. Wittgenstein was no scholar, having not read many other philosophers, though he certainly read Russell – a tale to which I return below – and he may have read some of Schopenhauer's views at the breakfast table in his Viennese home in his youth. In this way, fragments of Kant's thought must have slipped through the Schopenhauerian net into Wittgenstein's mind, and through Russell he would have acquired something of Hume's influence. Accordingly, you see Wittgenstein, in his notes *On Certainty*, reinventing some of the insights that Hume, Kant, and others had had about the assumptions we have to treat as undischargeable if we are to be able to claim knowledge or to be certain of anything. I mention this because at the very end of his life, therefore, Wittgenstein – having been committed twice over in two dramatically different ways to the idea that philosophy is a species of nonsense – was doing serious, central philosophy. It is the kind of work anyone would be required to learn about as a first-year undergraduate, and there was Wittgenstein taking it seriously at the end.

Here, however, is the game that Wittgenstein was playing in devising these two different philosophies in these two different phases of his philosophical life. You will remember in connection with the *Tractatus* that Wittgenstein said that the more important half of that book was its unwritten half. At the very end of the *Tractatus*, he says that what we cannot speak about we must be silent about: 'Whereof we cannot speak, thereof we must be silent.' This adverts to the point he insisted on, both in producing the *Tractatus* and at its end, namely that, unless a proposition is an actual or possible picture of a fact, the subject matter in question is something we cannot talk about, and that all the things that are much more important than what common sense and science address, namely questions of ethics, religion, and aesthetic value, are things we simply cannot talk about, but can only manifest our attitudes towards. He put this by saying that the world of a happy man is completely different from the world of the unhappy man – meaning that how you face the world, the stance that you take towards it, is in a holistic way expressive of what you are committed to in matters of ethics, aesthetics, and religion.

What lay behind this, the second of Wittgenstein's motivations (the first being of course to solve the problems of philosophy by understanding

A. C. Grayling

54

language), was that ethics and religion are far too important to talk about. Indeed, his desire was to protect them from what he regarded as the reductive encroachments of the natural and social sciences. He feared that, if science and psychology can explain why people hold certain ethical views and religious beliefs, the significance of them would be undermined. He did not himself express matters in quite the following way, but reductionism in philosophy and elsewhere is sometimes described as 'seeing nothing in the pearl but the disease of the oyster', thus taking away the value and beauty of things by giving natural-scientific or social-scientific explanations of them. In the sense that concerned Wittgenstein, reductionism would devalue the importance to individuals of (say) religion by explaining religious belief and experience solely in psychological terms. Wittgenstein was anxious to prevent this.

Thus, the game Wittgenstein was playing was to say that all the most important things we address in our living, and in our reflections on living, namely, our ethical and religious considerations, cannot bear discussion because discussion would be conducted in propositions. Yet they would be meaningless because these topics – these very important topics – are not the kind of thing which consist in actual and possible facts for those propositions to depict.

During the First World War, Wittgenstein read Tolstoy's version of the Gospels. When he later read the Gospels themselves he was very disappointed because they did not come up to Tolstoy's version of them. His interest in religion was serious; on three separate occasions he considered becoming a monk. It is evident from the nature of Wittgenstein's life that questions of personal morality, his own view, for example, of human relationships, were tormented. He was a man who never stayed in the same place for long, never spent more than two or three years in any one location, and very often took himself off to live in solitude. Evidently there was something painful and difficult in his private life. It may well have been because he was a homosexual at a time when homosexuality was regarded with abhorrence by many, and was criminalised by the law. If indeed he had strong religious feelings, these might further have complicated his inner life because of his sexuality.

It would explain a great deal about the unsaid and hidden motivations for his work that he wanted to make it possible for there to be either a barrier against the reductive encroachments of science on these precious matters, or – and here is the rather powerful result of the way he put things in his later philosophy – that it might be possible to justify and defend our attitudes in ethics and religion by saying that the discourses of ethics and religion constitute their own meaning, because they are a language game woven into a way of living, a form of life, therefore validating their own meanings, making sense to themselves because the game has its own constitutive rules. This way of valorising ethics and religion – especially the latter – has recommended itself to many who find that other ways of doing so too readily collapse in the face of rational inspection and because of a lack of objective evidence.

The Epistemological Contribution: Language, Knowledge, and Existence

There is yet more to be said about the way Wittgenstein uses the theory of meaning offered to us in the later philosophy. It connects to the following fact: Bertrand Russell was a great help to Wittgenstein throughout his life. He accepted him as a student before the First World War, helped him get his doctorate in 1930, helped him to get a research fellowship at Trinity, and then, when Moore retired from the Chair of Philosophy, helped him to the Chair of Philosophy there. Now it is a universal truth that, if *A* helps *B*, *B* will ever thereafter resent *A*. Certainly Wittgenstein seemed to bear a great deal of resentment towards Russell, not least because he thought, sometimes for not entirely incorrect reasons, that Russell's philosophy was insufficiently detailed and clear and that he made many philosophical mistakes – so he read Russell with the main intention of refuting him.

There were two aspects of Russell's views that came to inspire in Wittgenstein important contributions, which have remained a permanent possession of philosophy. Philosophers know them under the titles 'the rule-following considerations' and the 'private-language argument', and I shall explain briefly what they are.

One was this: Russell spent a few months in prison in the First World War, during the course of which he wrote a book called *Introduction to*

Mathematical Philosophy.[17] In this book Russell makes use of the notion of mathematical induction, the idea that the application of a rule will always guarantee the same outcome on each occasion. So, for example, the rule 'add one' requires you to give the successor of any given number and will always do so if applied correctly. This prompted in Wittgenstein's later philosophy a question: how can you be sure that applications of a rule will always have the same outcome? How do you know that a rule does not start to vary and produce different results far down some sequence of applications of it? Consider the rule 'add one'; maybe you get the immediate successor of any given number until you reach several billion, after which it suddenly starts to give you the second successor of any given number. How do you know – how can you guarantee – that a rule will always have the same outcome?

The other thought was this: again while Russell was in prison, he finished writing a series of lectures which he gave after his release. They came to be published under the title *Lectures on Logical Atomism.*[18] This was Russell's very informal and rather incompletely worked out version of views roughly similar to what Wittgenstein articulated in the *Tractatus*, because of course Russell and Wittgenstein had been discussing these matters together before the war – though during it they had travelled in rather different directions, one producing the austere architectonic of the *Tractatus* and the other the informal and sketchy version in *Lectures on Logical Atomism*. Fundamentally, the idea in both was very close. In Russell's version the world and language are of course structures also, but the bottommost layer of the world's structure is sense data (as Russell called them): atoms of information in one or other of our sensory modalities. These are the building blocks out of which we construct the physical objects that we encounter in empirical experience of the world. Physical objects are constructions out of the actual and possible sense experiences we have. Recall that, in the *Tractatus*, the atoms of the world

---

[17] Bertrand Russell, *Introduction to Mathematical Philosophy* (London: George Allen and Unwin, 1993 [1919]).

[18] Bertrand Russell (ed. J. G. Slater), *Collected Papers of Bertrand Russell. Vol. 8: The Philosophy of Logical Atomism and Other Essays, 1914–1919* (London: Allen and Unwin, 1986).

are objects, undefined, and that Wittgenstein said he did not know what they are, only that they must be there. Russell offered a candidate for the basic atoms of the world, namely the sense data we have. This is squarely in line with the empiricist tradition of thinking about how knowledge of the external world arises.

But the difficulty for Russell is that he made an assumption common to all epistemologists since Descartes. Descartes started from the contents of his own consciousness: I know what I am, he said, namely a thinking thing, and I know myself to be such because I have ideas, beliefs, and the like. From this internal perspective I have somehow to acquire certainty about what exists independently of me, outside my mind. Thus the Cartesian starting point is the internal content of one's states of consciousness. That is precisely where Russell starts likewise in his *Lectures on Logical Atomism*; from the little bits (in the computer sense) of sensation occurring in the inner space of experience, from which we have to infer an external world or somehow repose confidence in these internal states being genuinely representative of what is outside ourselves. This way of viewing the acquisition of empirical knowledge has been inescapably fertile in prompting scepticism, for nothing has ever been satisfactorily offered as a guarantor that private conscious states are by themselves true representations of anything external to themselves.

Wittgenstein recognised this as posing a major problem. If you were a kind of Robinson Crusoe figure and tried to talk about the inner states of your own consciousness, how could you do it? How could there be a language that was logically private to yourself in which you could discuss with yourself your experiences and what you take them to represent, so that from them you could construct knowledge of an external world? Here is an illustration of that problem. Suppose you are washed up on a desert island as a little baby before you have had any chance of exposure to language and somehow or other you grow up there safely, and one day a coconut falls on your head and gives you a headache. You say to yourself 'I'm going to call this sensation "headache".' Then, six months later, another coconut falls on your head and once again you have a headache, and you say to yourself 'Ah, now what did I call this the first time? Oh yes, "headache".' But how would you know that you had correctly used that word to denote the experience you are having on this later occasion?

Well, by memory, you would say: on this occasion you are remembering the baptismal occasion. But here is the problem; doing that is very like looking at something in *The Times* newspaper and thinking 'Good heavens, can that be so?' and rushing out to buy another copy of *The Times* to see whether what the first one says is true. So really you would, as a solo language user, never be in a position to know whether you are using terms with the same meaning as on an earlier occasion of use. You could only do this as a member of a linguistic community which could check and control your uses of expressions. To be a language user, therefore, requires you to be a member of a rule-following community of speakers who, together, constitute the meanings of the expressions they use and govern their uses of those expressions.

This is the important point Wittgenstein raised about the impossibility of a logically private language. Of course there could be a *contingently* private one, like the code in Pepys' diary, but the point about such a thing is that it can be translated into a public language and so it is not *logically* private. But if there cannot be a logically private language then there cannot be a Cartesian starting point for the theory of knowledge.

That was the big contribution Wittgenstein made in reacting to Russell's views. You can see the connection between the rule-following considerations and the private-language argument; they are in a way the reverse and obverse of the same coin, because the reason why a language cannot be logically private is that language is a rule-governed activity, and rules can be followed only in a public setting.

Conclusion

This, in summary and non-technical terms, is my account of Wittgenstein on games, and that is my account of the game Wittgenstein was at the same time playing: namely to protect the things he regarded as important – ethics and religion – from the encroachments of reductive scientific attitudes. He did this in his early work by saying that we cannot talk of ethical and religious matters, and he did it in the later work by saying that ethical and religious discourses validate themselves because they give themselves their meaning in the language games and forms of life they constitute.

As a concluding thought: there is some interest and significance in this last point for us today, because we live in a time of bad-tempered disagreement between people who have a religious outlook and people who do not. There are arguments about the degree to which religious assertions and beliefs have content, and whether the belief that people have in their being contentful should be one that they keep to themselves, and to what extent, if any, that should be permitted a privileged place in the public sphere and in public policy debates. What Wittgenstein had to say touches on the great question of whether religious talk makes sense. That is where the crux of the debate between the two sides in this bad-tempered argument lies: is religious discourse literally meaningful or is it not? Wittgenstein's 'games' idea offers comfort to those who think that religious talk is meaningful. For those who think that it is not meaningful, the metaphor of 'games' seems more than peculiarly apt.

4 Games in Sports

DAVID BRAILSFORD

As the Team Principal of Team Sky and having run the British Olympic Cycling Team for several years, I have been very, very lucky to see how UK cycling, as a sport, has developed from a lowly competitor to a world-beater. During that journey, I have been very interested in the games that I have played: there is the essential game of strategy or planning, and there is a game of tactics and execution; both are very important. However, the most important game in sports is the game of the mind and how the mind is used to make the difference between elite performers. I often use models and methodologies to think about games within sport, and I draw these models from my experiences within cycling.

To begin a brief history of my involvement in cycling, my father was a very keen mountaineering cyclist. He used to get all his gear and lycra and, as a teenager, I was horrified by the sight of him. I used to say to him 'Look dad, if I'm standing on the street corner of our little village with all my friends and you come past looking like that, you can wave as much as you want, I am not waving back.' Cycling was not a main-stream sport at the time; it was a very minor sport. But, at about 19 years old, I began to pursue cycling and decided that I was going to win the Tour de France. I packed my bike in a cardboard box, got a single ticket from Bangor, North Wales to Grenoble, and off I went in pursuit of my dreams – rather naïvely so. There was no Channel Tunnel, of course, at the time, there was no Internet, and there were no mobile phones. I vividly remember getting on the train at Calais to go to Paris and then to Grenoble and the woman came along with the coffee trolley and asked me 'Une tasse de café?' I thought 'Oh my God', I did not understand a word; and it was only then that it dawned on me that the whole thing was going to be a bit more of a challenge than I had realised.

To cut a long story short, I got to Grenoble, managed to find the end of a professional cycling race, and chose the team with the coolest-looking kit. I went up to them and asked whether I could join their team and race for them. They passed me round from team to team, and a man called Pierre Rivory very kindly said that if I came to Saint Etienne on a Wednesday morning I could train with the team and take it from there. I suspect they thought they would never see me again. But I made my way to Saint Etienne on a Wednesday morning and spent three years racing with them – rather unsuccessfully, I hasten to add. During that time I realised that despite being a real trier – I did my absolute best – I was not going to win the Tour de France after all. But what I did find, thankfully, was an absolute passion for anything that would inform me from a sports science, technological, nutritional, or psychological perspective how to improve human performance and how I could get better as a cyclist. I found my passion in that journey.

This was not what I expected to find. I came back to the UK and went to university to take a degree in sports science and psychology, which I absolutely loved. I just would not leave the lecturers alone, I wanted to learn more and more and more – I think I was a bit of a pest in the end, they were glad to see the back of me. I then wanted to follow a career in sports psychology, but at the time it was difficult to really see how that path might develop. Sports psychology was not truly accepted in professional sports at the time, particularly in football. I therefore went to Sheffield Business School to do an MBA, though I kept a foot in, if you like, with the British Cycling Team. Thankfully, in 1997 National Lottery funding came along, which changed the face of sport in this country for all of the summer and winter Olympic sports. And that, of sorts, is my academic background and my entry to cycling.

In this chapter, I will demonstrate how games in sport have applied to my life in cycling. As part of this, I will demonstrate some of the models and methodologies we have used within Team Sky and British Cycling to play these games and improve our riders.

The Ascent of British Cycling

When thinking about games and playing games, it is interesting to ask why a nation that is good at playing one game may not be good at another, and to interrogate the cultural aspects behind this. There are eighteen gold medals available in cycling disciplines in the Olympic Games. Prior to 2000, Great Britain had won only one gold medal in cycling over 76 years of effort. That meant that we were not very good at this game. This is interesting, given our population and demographics, which are not overly different from those of the nations who are very good at cycling and were winning medals. It was not that that we lacked the people, the talent, or the ability – but something was amiss. The only medal that we had won was by Chris Boardman in Barcelona; he won the individual pursuit on a fantastic, very futuristic bike that had a monocoque frame made by Lotus, the car manufacturer. It was a beautiful piece of machinery. If you ever come across Chris Boardman, do tell him that it was the bike, not him; he absolutely hates that.

The Atlanta Games were our low point in Olympic history, we were twenty-fifth in the overall medal table – Pinsent and Redgrave won gold medals, which for a nation of our size was pretty dire. You have to think, how did that happen? What was going on in that era? When we went to Sydney, the mentality of a lot of British Olympians at the time was that the top of the mountain was getting selected for the team. Becoming part of the Olympic team was the goal; the aspiration of winning and the belief that you could win was not truly there. That mentality represented where we were as a nation in sport in the 1990s: the gallant losers, getting close but falling at the last hurdle.

So the question is, what happened? Of course, the catalyst was the advent of National Lottery funding. It was, in my view, the single biggest factor in changing the face of how we currently perform in most games and certainly in the Olympic Games. The only thing the Lottery was interested in was how we could move up the medal table as a nation. Lottery funding gave us the opportunity to develop a 12-year plan, and my predecessor – the brilliant, visionary Peter Keen – wrote a terrific plan for cycling that received significant funding.

Thankfully, cycling moved on and we started to really concentrate. Jason Queally, on the first night of the Sydney Olympic Games, got up and delivered a masterful performance that nobody had expected. He had switched to track cycling from water polo, delivering our first gold medal; and that was it, from that moment we had belief. There has been a shift in mindsets over the last 15–17 years more broadly. From then, we started to build and move forward. We went to Beijing knowing that we were in a good position; illness was the only thing that could potentially hold us back – we did not want to get ill. There were ten medals available in the track disciplines, and we managed to win seven of those ten gold medals, with another couple of silvers and bronzes to boot. The greatest part of that was the overall medal table: if British cycling had declared independence and become an Olympic nation in its own right, we would have been higher up the table than France. That gave me a lot of satisfaction, I have to say.

Next, we had the challenge of sustaining success. Part of the journey in this chapter is about trying to create success and becoming successful. The next challenge – a different challenge altogether – is sustaining success. In my experience, these are two very separate things. When the London Games were announced it was a terrifically exciting time for us, but, having gone to different continents to be at the Olympic Games, then being able to drive down the M6 for the Games brought its own challenges. We were under pressure and we felt it. We felt that, if we did not perform to the same level in London, in front of a home crowd, as we had in Beijing, we would be deemed failures. So, it was a very different Games, necessitating a very different approach. Thankfully, we managed to deliver.

Taking that same methodology, we then felt that if you really want to be successful in a sport you have to think about the biggest possible achievement in that sport and try to win it. I felt passionately about the fact that no British rider had ever won the Tour de France. We had conquered the Olympics, so, next, we should try to get a British person to win the Tour de France – clean, I hasten to add. It became a fantastic new project for us, and 2012 was a brilliant year in that respect: a terrific 2012 Olympic Games and Bradley Wiggins winning the Tour de France.

Progression, Not Perfection

What lies behind this ability to improve and achieve, from a sports and games perspective? That is the question I am exploring in this chapter. In my experience with cycling, we have stuck to some core principles in approaching these challenges of achievement. These core principles remain useful today, and I always map out our game strategy in terms of these pillars. The first set of pillars is about strategy, and performance strategy in particular. The second, probably the most important pillar – and certainly the pillar I spend most time thinking about – is the human mind, human beings, and, more recently, mindset culture. Culture is currently fashionable in sport; many books have been written about the New Zealand All Blacks' culture in rugby, and globally many teams focus on getting the right culture to achieve results. Get the culture right and the wins will follow. But I am not sure I agree with this; I believe more that you do what you can to win, and build the culture from that. However, it is a chicken-and-egg scenario over which there is a lot of current debate.

Third, and I'm not sure whether I am pleased with having coined the term because I get asked about it wherever I go, are 'marginal gains'. When we were a million miles away from the Olympic podium and looked at the top of that mountain, people did not really believe that we could just go and win. We felt that it was psychologically important to try and break the challenge into something where we could take control. I am a believer in the idea of progression rather than perfection. In an Olympic final, for instance, the athletes have been preparing most of their lives for these very short performances, so they want perfection. However, as humans rather than robots, even the best sportspeople are fallible and will make mistakes. They will have anxiety, just like the rest of us. I think that aiming for a perfect performance is not optimal for humans, we should instead think about progression. When asked, even the very best athletes think they could get a little bit better and find that idea quite motivating by its very nature, whereas wanting a perfect performance is quite intimidating. Consequently, I am a big believer in progression rather than perfection, and this is where the marginal gains theory came from. We broke down every element possible in relation to cycling per-formance and asked ourselves – forgetting, for the moment, the top of the mountain – could we improve all these areas by a small amount? Everyone would say that we could, that it was not a problem. Suddenly, therefore, we had an engaged group and felt that we could improve. We started making progress and the perfection could come later. I will elab-orate further on the marginal gains idea below. First, I will examine each pillar of the core principles in turn.

Performance Strategy

When you think about a strategy, certainly in sport and particularly at the end of a season looking forward to the next season, there is a tendency or cognitive bias towards focusing on the limitations that you currently have. In other words, you become stuck by the limitations that you see or feel at that moment in time. Within cycling, we therefore try to work in a way that lets us forget the present and instead go into the future to visualise what it is like to achieve the performance that we want to achieve. The first step is to work back from the goal. We call this process 'the analysis of the demands of the event'. If you want to win a game, you need to know how to win it before you start; therefore we spend a massive amount of time analysing the demands of the event. In cycling we now have access to better telemetry, making analysis and measurement easier. We can better understand the physics of the sport, and we can take better physiological measurements too. We accumulate a large amount of data about what we can try to achieve and what it will take to win, but we also try to look at the situation more broadly too.

Take, as an example, the Tour de France on a normal race day. We have a recovery protocol for racing, but the organisation says that a rider has to go to a press conference after the race. So instead of having an optimal recovery protocol, which we worked hard to develop, if you become the leader of the race you are handicapped because you have to go and sit in a hot tent to do the press conference. To address this, we have an individual car that looks after the race leader and the rest of the team are bussed back to their hotel because this individual has to stay and wait. The press are therefore a real challenge for riders. Given the history of cycling, this is an increasing intrusion as the more successful we become the more questions the press ask in terms of whether or not the team is cheating. This is challenging and difficult to deal with when you are not cheating.

Ultimately, however, if we want to win the Tour de France and create a document called 'what it's going to take to win', we need to be able to deal with this type of situation, even though it may not be a situation we immediately expect when thinking about the Tour de France. It is, however, an absolute must in terms of strategising and working towards becoming effective to plan for these types of situations.

We also have to think about other aspects of the Tour. For example, in 2016 there was a renewed focus on individual time trialling, whereas the focus had been on team time trialling in 2015. In 2011 and 2012 Chris Hoy and Bradley Wiggins were doing fantastically in time trials, with both riding eleven competitive time trials in 2011 and around thirteen in 2012. In the following two seasons, however, they rode only four and five time trials, and so were less prepared. In 2016, the individual time trials in the Tour de France fell the day after the Mont Ventoux stage. We knew that there would be time gains or losses in the time trial and that the riders would have to put a lot of effort into the Mont Ventoux stage. From a strategic point of view, the question then became 'How do we want to approach this?' There was also a subsequent time trial later in the race, again just as important. When considering the climbing efforts coming before and after it, there was again a strategic decision to be made in and around how to approach those particular aspects of the race.

There are multiple other factors that need to feed into a performance strategy. For example, we do not just ride on flat roads. There are likely to be a few cobbles, and when these get wet it is a new challenge altogether. Of course, there is also an equipment challenge, and the riders have to train on the cobbles to get used to them if you do not want to lose any time. In the Tour of Italy there is extreme heat and extreme cold, which proves a real challenge for us. So that also has to form part of the strategy. Further, in every professional cycling race, somebody will crash: we have never finished a Tour de France with a full complement of riders. We know that we are likely to finish the race without the full nine starters, but we do not know whom it is that we will lose or when we are going to lose them. That is another challenge from a strategic point of view. When a rider crashes, their instinct is to get straight back on the bike and carry on. Of course, from a medical point of view, if they have fallen and hit their head or landed on their head there is a discussion about concussion, and this is a significant point at the moment, particularly in rugby and American football. We have to be very careful in assessing the riders for concussion and tell them not to continue if we suspect they have concussion, even if the rider wants to continue. That is not easy, and results in a conflict between the medical staff, who want to take care of the riders, and the ex-riders, coaches, and everyone else who wants them to just get back on the bike. We therefore need a policy on this to try and deliver our strategy.

The less obvious part of a strategy – though it should be more obvious – is knowing the rules. As an example, there is a rule in cycling on the flatter stages that if you puncture or have a mechanical problem within 3 km of the finish you are awarded the same time as the lead rider. During the Tour of Italy in 2015, Richie Porte was our lead rider. Richie punctured with about 3.3 km to the finish, which was very unfortunate. Another Australian also punctured alongside Richie. The Jayco rider was a domestique, a helper rider, whereas Richie was trying to win the race. He instinctively took his wheel out, threw Richie's to one side and put his own wheel on Richie's bike, said 'There you go, mate', and pushed him off. Most people who saw that on television, and all the different sports directors, who communicate frequently, thought 'Wow!' In sport and in games, sportsmanship does come into the equation. That was sportsmanship. It was instinctive; there was no time to reflect or calculate or think of the consequences. It was a very positive gesture – though not according to the referees, unfortunately. We received a phone call two hours after the race had finished, informing us that Richie had been penalised three minutes and that his race was over. So, as part of the overall strategy, knowing the rules is very, very important.

When thinking about games in sport, it is interesting that we have an increasing number of -ologists in sports; every -ologist has a helper who is also an -ologist, and it is growing all the time. What this allows us to do is to really zone in on that part of our strategising and planning. One of the key things, for me, is to spend time and effort understanding what winning looks like. As a team, we spend more time analysing the demands of the event than anybody else, and it gives us a clear goal; it helps.

Chimps on Bikes: The Human Mind, Human Beings, and Mindset

Next, I turn to discussion of the human being and the human mind. From a personal perspective, this is the area that has influenced me most and in which I have gained the most insight over my career. All sportspeople are human, so it seems interesting to me to gather as much information as possible about human beings, particularly through sports psychology.

In 2000, I had a situation with one of our riders who had a clinical issue beyond our expertise. It transpired that our team doctor had been a student under another doctor at the Sheffield School of Medicine, Dr Steve Peters. Dr Peters is a surgeon, a mathematician, and a forensic psychiatrist; he's a unique character. He was the undergraduate dean at Sheffield School of Medicine, and also worked at Rampton with mass murderers and psychopaths. Dr Peters came and worked with our rider – he fitted into the team brilliantly. There was a miraculous turn around, allowing this rider to perform very well – he medalled at the Commonwealth Games in Manchester in 2002.

I was blown away by this, and was really interested in what Dr Peters had done, so I asked him to meet with me. Dr Peters uses a model that allows you to identify a person's belief system and so do x to achieve y result. This approach chimed with me, and was something I was keen to learn more about. I therefore managed to persuade him to come and work for British Cycling. The questions turned from cycling to humans and questions about humans playing games and sports. How do we optimise the performance of a human being? What does a high-performance environment or a high-learning environment look like? Dr Peters created a model for this that we have used to great effect. The brilliance of Dr Peters' work is that he takes something complicated and complex, like the brain, and creates a simple model that allows us to develop systems and interventions, a way of thinking, and a language and culture around it. It simplifies something very complicated, but it has been very powerful. I will now explain this model and its application to British Cycling and Team Sky.

The brain has many different parts, and our behaviour is dictated by which part of the brain is thinking for us. It is, of course, a system; different parts of the brain are always interacting. In cycling, we are interested in the limbic system, which is the emotional part of the brain – your chimp, to put it simply. The limbic system does not use logic or think about consequences, and is very individual. The limbic system can be irrational; it is where our fight or flight mechanism resides. We have no choice over what our limbic systems are like. In sports it is interesting to delve into the question of why we sometimes act involuntarily in a way we do not necessarily want to – the reason for this is that our limbic system, our emotional reactions, are controlling us. If you overrode the limbic system you would perhaps act more on logic and think about consequences. You could be more analytical and control your impulses.

When driving a car for the very first time, you think deeply about how to drive and about all the different components like the gear stick, the clutch, how to manoeuvre, and so on. But, two or three months later, you can happily drive along without thinking of these things: your parietal lobe is driving the car for you. What is interesting for sport is that, if we could allow that part of the brain, uninterrupted, to play a sport, it would create optimal performance. We term this the zone or flow – a term used a lot in sport.

These two systems – the limbic and parietal – within the brain are constantly battling one another. As an example, consider a situation in which you receive an email that you are unhappy about and send an impulsive reply despite having hovered for a second over the 'send' button. Ten minutes later you will find yourself regretting sending that email, but your limbic system sent it for you. This explains a lot within sports and gaming contexts. When the emotional part of your brain is controlling your behaviour, you become erratic and illogical. This can also affect the thoughts athletes have, such as 'What if I lose?' or 'Oh no, that rider looks fitter than last time.' Athletes have emotional responses, and we have to deal with this in gaming scenarios. In particular, this is important when making decisions under pressure. If you allow the limbic system to make those decisions for you, they are likely to be irrational, illogical, and certainly not optimal. Equally, however, there is a danger in overthinking, where the frontal lobe and the limbic system have a long dialogue. Sport requires instinctive decision-making.

We therefore work intensively with our riders on how to make decisions in high-pressure environments; on how to approach challenges like competitive anxiety before a competition. The aim is for the parietal lobe to play the sport and to occupy the other parts of the brain to keep them out of the way, as it were. For example, take England's less-than-illustrious history of penalty-taking against Germany in football. If our footballers could take their penalties with only their parietal lobe, we would win every time. So, what happens in that scenario instead? Why is the brain instructing muscles to behave in a certain way to kick a football affected by being in a stadium of 80,000 people or a perception of the consequences of kicking the ball? It is a purely nervous moment, but it does have an impact. That is what we attempt to address with this model.

We have also used this model effectively in cycling in terms of communication. Conflict is inevitable within groups of people who spend significant amounts of time together in high-pressure environments. I used to believe I would have brilliantly harmonious teams with no problems, but over time you realise that with groups of people it is nearly impossible not to have conflict. So, when somebody gets agitated in British Cycling or Team Sky, they come to me and I tell them to let their chimp out. They then rant and rave and eventually run out of steam. You know that this is irrational, it is just their limbic system, and I pay no attention to it. Once they have finished letting off steam we can look more rationally at the situation and quite often the person then feels much better; it was often simply about wanting to alert me to something.

There is a danger, of course, that in this type of situation my own limbic system might kick in – two chimps going at each other. This would be a problem. You often see this happen in sport when, for example, football managers get angry and gesticulate at the side of the pitch. It shows that many of the decisions made in sport are purely emotional decisions. Another example is when Jason Queally won the kilometre time trial at the Sydney Olympic Games. A French rider, Arnaud Tournant, was the favourite; he was reigning world champion, a fantastic athlete, almost unbeatable round the track. He was going to win the Olympic gold and everyone else was turning up to ride for silver and bronze. Jason was fifth-to-last to do his ride and did a fantastic time – it was the winning time in the end. But in between Jason finishing and Arnaud Tournant taking his ride, he huddled with his coaches and they decided to put on a bigger chain ring, more teeth, and a bigger sprocket than he had ever ridden with before in competition. When you analyse that decision, it is purely Arnaud Tournant's limbic system reacting. He then took his ride and went out fast, getting up to a cracking speed but within half a lap of the finish he could have got off the bike and walked faster than he was cycling; it was inevitable. This is a great example of the sort of emotional hijack that can occur in sport if we are not careful.

The model that we use to get the best out of humans in a high-performance environment is very simple. It is called the Core Principle and is focused on how to achieve personal excellence. Developing and using models like these is very important to me because I have never won a bike race – I never will now, unfortunately – and yet it is the single thing that I get judged on. My entire career is judged on what other people do on the bikes, not what I do. That is why this way of thinking and these models are so important, because I have to learn quickly and become an expert in helping other people to be the best that they can.

The first step is commitment, or hunger, if you like. Commitment minus distractions provides you with the opportunity for success. Somebody saying they are committed to winning a gold medal is not the type of commitment I mean, it is more deep-seated than that. When we talk about commitment it is more along the lines of trying to understand the demands of the event and what it is going to take. You take an audit of where an individual is now and have a clear idea of the gap between that and the demands of the event. You then assess whether this is a bridgeable gap or not and, if it is bridgeable, put the appropriate interventions in place to try to move from *A* to *B*.

The next step is to map out in an individual's day-to-day life what delivering that achievement looks like. For a lot of our riders there is a need to be hungry all the time, they have to be committed all the time. To achieve the power ratios they are now capable of requires a lot of sacrifice. Riders have to be away from their families, they have to train huge numbers of hours, and have many other demands placed upon them. You therefore map out the sacrifices they need to make. What I am interested in is their commitment and their willingness to make those sacrifices, because that will have a greater impact on the outcome than anything else. Riders need to be intrinsically driven and if a rider is not intrinsically driven towards achieving there is little that can be done: I cannot work with them unless there is at least a little flicker of a flame. I cannot reach inside somebody and turn on that switch. Thus, the first, and crucially important, assessment is around commitment and talent.

I use something I call the 'hunger index', in which – at the end of every year, at the start of a season, and periodically afterwards – I put myself in our riders' position and attempt to see from their perspective. I try to understand their family situation and their position within their contract (it comes as no surprise that there is a strong correlation between a rider being on their final year of a contract and good performance), as well as other factors. Doing this allows me to better assess each rider's drive. But, as humans, nothing is static, and there is never a constant level of intensity, meaning that the success or not of the previous season as well as family situations can have a dynamic impact on a rider's hunger and drive. The preceding season having been good or bad is a particularly interesting case. People can change behaviours when the reward or suffering is great enough, but when it is not great enough people tend not to change their behaviour and instead stick to the same patterns. Sticking with old behaviours can be damaging, though we are creatures of habit, and we often need a catalyst to shift us out of those behaviours.

The next biggest challenge is the idea of ownership. The idea behind this is that humans will perform better if we have ownership over what we are doing; we like to have an input into what we do. We therefore decided to take the crowns off the coaches' heads, as it were, and transfer those crowns to the riders. We were going to make them the kings and queens of their own destiny. I nearly had mutiny on my hands because the coaches did not like this; they perceived it as giving away control and power. But that was not the aim. By prompting ownership and promoting the belief among the riders that what they were doing would work and that they could negotiate their training programmes, we found a method that worked better. A one-size-fits-all training approach will never work. If you want to be successful and win games, you have to recognise individuality and personalities. The key is finding the right solution for a particular person to develop. I believe that taking the time and the effort to really understand how individuals work and give them some ownership over what they do, asking them how they want to be supported, asking them how I can help, and providing the support that they need is a hugely powerful act.

Some people, when given ownership, give it straight back to you and just want to be told what to do. There is no problem with that, but the fact that you offered the ownership is enough. Others want total ownership: they want to write their own training programmes and do not want to do what everyone else is doing. Whether we believe in the programmes they devise or not, whether they are optimal or not, it is a mistake for us to refuse this and impose our own programmes as coaches. Fundamentally, the power of a rider believing they can improve is one of the greatest assets you can have, in my opinion, for reaching optimum performance. If you do not believe in what you are doing, the opportunity to be successful and achieve personal excellence is greatly diminished. This makes ownership very important.

Consequently, if you give people ownership then they must accept responsibility and accountability. There is a big difference between personal and professional behaviour, and in sports teams, as professionals, we are bound by codes of conduct. It is a professional environment with professional codes of conduct. When the team spends a month at a time preparing for the bigger races, it can become very difficult not to allow the group to become more of a personal environment where the team members operate on a personal level. For example, we do not have favour-based relationships in the team because this would not be professional, it would be personal. We have certain rules and we educate our riders in the difference between personal and professional in terms of the relationships we have to have. We also ask our riders to write their own rules, meaning that, from a responsibility and accountability point of view, in the light of our emphasis on ownership, we try not to police the riders. Interestingly, the rules the riders write for themselves are probably stricter than the rules we would write for them. Ultimately, when it comes to managing a group and a team in a sporting environment, it is powerful to allow the riders to decide what their own rules are rather than having rules dictated to them. This leads to personal excellence.

Another issue within sport and games is the idea of the team and the individual – in other words the concept of 'no I in team' or the great team player. It is almost folklore: the individual who is the fantastic team player. I would challenge this, however, and, though it is based solely on my own experience, I believe that the 'no I in team' aphorism is a complete load of rubbish. We all think about ourselves first and foremost. Yet I would argue that there is a continuum; we are in a constant state of flux between thinking about ourselves and self-sacrifice. It is interesting to understand how this works within each individual rider. I may not agree with their perspective, but that is irrelevant; I simply need to understand what they are thinking in order to get the best out of them and to manage the situation.

Cycling is interesting because the rider who wins cannot win unless everybody else in that team sacrifices their own opportunity day after day for that rider to win. Even then, the team doesn't win, the individual does. Nevertheless, in, for example, a Tour de France team there will always be some riders who are never going to win themselves but are quite happy to sacrifice themselves because they want to be part of a winning Tour de France team. Mark Cavendish in 2012, for example, could win individual stages of the Tour de France but he could not have won overall. So for him to contribute to the Tour de France winning team, he had to sacrifice some of his own opportunities for the team's benefit. That is a big ask.

On another level, you may have two riders, both of whom think they can win. Using this model, it can be easy for us, as onlookers, to tell them to be team players and urge them to be professional, but that is not how the two riders will see the situation. In their eyes the sacrifice that needs to be made is not proportionate. When this happens I try to use the model to fathom how to approach such challenging situations. As an example, we had a similar situation in the Alps in 2012 between Chris Froome and Bradley Wiggins. Bradley was the strongest rider in the race, and Chris wanted to finish second, feeling totally within his rights to do so as it seemed clear Bradley would not lose the race. Chris was helping shepherd Bradley up a hill, Bradley dropped back, and so Chris went off and attacked. Then, over the radio, Chris was told 'Oh Froomie, I hope you got Bradley's permission to do that', making Chris stop dead in the road and come back. I think he realised the consequence and came back round to the idea of self-sacrifice for the team. Of course, when we reached the top of that mountain, all hell broke lose. It was an interesting scenario to manage. People often suggested simply getting them together in a room and knocking their heads together to knock some sense into them. But I did not think that was the right approach. I think you cannot change someone's personality and have to instead work with personalities. As you have to work with what you have, you need to try to understand it. That is what I therefore tried to do in that situation.

This led to a difficult situation the following year during the team selection for the Tour de France. Selecting the optimal team for the race is a challenging decision. A significant body of literature focuses on the importance of team harmony, but I do not believe that team harmony is the 'big thing' that we should always be seeking. Personally, I think goal harmony is most important. Therefore, the critical element in composing a team for a game scenario is to ensure that everyone in the team is aligned behind the goal – a goal harmony. Aiming only for team harmony misses a trick. Ultimately, it is easy to make popular decisions when selecting a team, but it is emotionally difficult to make the decisions that you believe will create a winning team. This is an important aspect of games and gaming: are you playing to take part or playing to win? I play to win – I always have, and hope I always will.

When devising your strategy, there are a million different approaches to playing to win, and this requires significant thought. So, to return to the difficult situation I faced selecting a Tour de France team, I began to think that I might not select Bradley Wiggins for the team, despite the fact that he had won the Tour the previous year and had been the first British rider to win. That was not a nice situation to be faced with, but I had to think about the most likely opportunities we had for winning the race. When I worked back from that and analysed the demands of the race, the solution became apparent. That is not to say it made the solution any more palatable, however. I deliberated for a long time, questioning whether I should instead make the 'popular' decision. It was certainly one of the toughest episodes of my career, and I am sure that some people would argue that I made the wrong decision. Nevertheless, my decision was ultimately based on this method of thinking.

I am sure that if I asked you, the reader, to write down the recipe for a fantastic team and a fantastic team culture, you would create something just as good as anyone else in the world. I think we all have knowledge about how great teams work and what to do, but it is the execution of that knowledge that makes the difference between success and failure. Most businesses will have their culture written up somewhere on the wall, using words like 'respect', 'honesty', and 'integrity'. But does this really filter into someone's behaviour at 11.30 pm on a wet night when the mechanics are outside having a conflict? Probably not.

In Team Sky and British Cycling we therefore attempt to identify within small groups what difference there is when a great culture exists.

We have thirty riders and a team of ninety overall, which we constantly separate into smaller groups. Sometimes, within those smaller groups, there is a culture that just feels fantastic. At other times, a particular group just will not feel right. We therefore analyse which behaviours promote good cultures and bad cultures; we analyse the 'winning behaviours', as it were. Once we have identified those winning behaviours, we can start to work towards them. Interestingly, a lot of literature focuses on the positive aspects of a culture rather than on losing behaviours. Yet losing behaviours are very damaging within a team culture and, for me, have an exponentially greater impact on a team culture than winning behaviours do. For example, moaning is a losing behaviour. One person being upset and moaning about it can quickly spiral among the team and turn into a damaging negative undercurrent.

We therefore identified what winning and losing behaviours were, and categorised them. We then developed a Team Sky app, the idea for which came from seeing happy faces and sad faces to determine levels of service and satisfaction almost everywhere, from airports to banks. During a race I ask our staff to use the app to register where on the scale (of happy to unhappy faces) they currently think they are. It asks whether they think they are having a net positive or negative (winning or losing) effect on the team. I try not to be too 'Big Brother' about using the app, but I do monitor it very closely so that when I see that someone feels they are having a negative effect I can try to help, support, and understand that person, as well as giving them an opportunity to have a more positive effect. Our cultural lens, if you like, is therefore about doing little things every day, and this does have a significant impact on people's behaviour.

Marginal Gains

Finally, and briefly, I turn to marginal gains and how we use them within cycling. As I noted above, we decided to aim for progression through small steps rather than perfection within sporting performance. It is rare in sport that a big step change in performance can be achieved, though it can happen through innovation. Generally, I believe you have to constantly look for innovation at the same time as making small steps. You must have both marginal gains and innovation together, and this is the philosophy we have adopted for continuous improvement. Ultimately, you produce a series of small gains, which, added together, enable a significant step forward.

For example, when racing the Tour de France, we stayed in at least twenty different hotels. This means we have no idea what the beds will be like, or what the hotels will be like overall. We can do some reconnaissance, but ultimately we have little control because the organisers dictate where the team stays. But if you have nine riders who want to race for three weeks, why would you allow them to sleep on a random mattress, with a random pillow, and risk having their posture and spine misaligned? We have osteopaths and physiotherapists constantly working with our riders and keeping them straight, which would be wasted if they were to then sleep on a random bed. So, we have a small team who go ahead to every hotel and remove the beds, clean the rooms, then put our own beds in so that each rider has their own mattress, pillows, and sheets. The riders therefore get to sleep in the same position for every night of the big races. Is that going to win you the Tour de France? No, it is not. Is it going to contribute in a small way? I certainly think so.

We work a lot on diet too. Our approach to diet and nutrition has changed dramatically over the last 10 years. What we used to believe was an optimal nutrition strategy 10 years ago looks very different from what we would consider optimal now. The challenge is to retain muscle mass and become as lean as possible to maintain power, while dropping weight. The process of timing when to eat and eating the right foods in relation to training and racing has become highly detailed. We now have eating plans which focus on the particular training or race that is coming up in the next few days, and the plans are constantly monitored. The difference is as stark as between night and day when you compare our approach to nutrition with the carb-loading approach taken in the not-too-distant past.

There are also opportunities to find marginal gains with equipment, and this can be a lot of fun. Aerodynamics is massively important, but you cannot take it too far as marginal gains always need to be related to importance. For example, we have a rider – Ed Clancy – whose role in team pursuit is to start as the first rider. He will go out and accelerate to 64 kmph as quickly as he can with the rest of the team behind him.

It used to be the case that Ed could accelerate for three-quarters of the first lap and then swing out of the way, leaving the next rider to do what we call the 'pick up' to take the team to optimal speed before this second rider would then get in the back and the remaining riders would go off. However, eventually we decided on a better approach that meant we would not lose the bike length we had been losing every time the team changes the rider in front. Instead, we trained Ed to do a lap and a quarter start and to get us up to the optimal speed more quickly but still be able to keep with the other riders for the full 3 km. Basically, no one else had done this because it was assumed that a rider would be so tired after the initial sprint they would be unable to maintain the pace until the end. Ed became the first person to prove that wrong.

Most professional cyclists break a collarbone, often both of them, at least once. Ed has broad shoulders, which prompted our aerodynamicists to suggest that taking just an inch out of each of Ed's collarbones would make a difference. We hypothesised that nobody on the planet could get close to him on the track if he could just get his shoulders in. We therefore suggested this to Ed, and, after sleeping on it, he agreed to give it a shot. At that point, a doctor in the room said 'Stop, what the hell are you lot thinking about? Stop right there.' We all shook our heads and thought 'Oh, blimey.' So, we need to be careful not to take these marginal gains too far.

Conclusion

This account should have given you an overview of, and an insight into, how we play games – both psychologically and physically – through some of my experiences over the last 20 years. I hope that you have enjoyed reading this chapter and have taken something away from it.

5 Losing the 'New Great Game'

FRANK LEDWIDGE

There are few more pleasant places in the world than Cambridge. In stark contrast, there are few worse places than Basra and Helmand, for many years home to British soldiers. They, and their NATO allies – many of whom were decent, well-trained professionals trying to make the best of a bad situation – operated under the ever-present threat of attack, aware that many of those in Basra and Helmand were intent on killing them. Six hundred and thirty-five of these personnel never made it back home alive.[1] Thousands more were seriously injured; yet more suffer severe psychological consequences.

I do not propose here to discuss what it was all for. Nor will I look in any detail at the operational techniques or tactics used, or consider the violence visited upon civilians – including the thousands of Afghans and Iraqis who are now dead, injured, or sick as a result of our actions. We all have a stake in these wars. The Helmand campaign alone cost in the region of £37 billion.[2] These were savage contests; wars where, in the words of Clausewitz, even the easiest thing was indeed very hard.

All that notwithstanding, these were, I contend, failed operations. I will be looking at three of a much larger set of factors that led to failures: (a) generalship and strategy, (b) the closely linked issue of accountability, and (c) intelligence. This chapter will not discuss a lack of helicopters or insufficient numbers – neither of which played a part in strategic failure. Nor will the focus be placed upon the role of politicians, however significant. I will explore what to many may seem, and really was, a chaotic six-monthly changeover of the entire force in each conflict, producing a series

---

[1] Based on 456 UK Ministry of Defence fatalities in Afghanistan and 179 in Iraq between the start of each intervention and the time of this book going to press. See www.gov.uk/government/fields-of-operation/afghanistan and www.gov.uk/government/fields-of-operation/iraq [accessed 30 August 2018].

[2] See Frank Ledwidge, *Investment in Blood* (New Haven, CT: Yale University Press, 2013) for a full estimate.

of biannual wars where each commander had a signature fiercely titled operation, his 'trade-test', as one general called it, rather than a coherent strategically integrated whole. This was driven by the requirements of the structure of the army, rather than any kind of rational approach. Each campaign ran a similar course, namely initial tactical success, overstretch, a bailout by the US armed forces, and, finally, withdrawal.

The Deployment to Iraq

In 2003, the UK provided no fewer than one-quarter of the combat troops involved in the invasion of Iraq. Choosing from a set of three options, senior military officers chose the plan which involved the largest possible involvement of British military power, involving the taking and occupation of Iraq's richest province. No consideration was given to the responsibilities in law and fact which necessarily followed. Both the fourth Geneva Convention of 1947 and the Hague Convention of 1907 provide a detailed series of obligations for occupying powers. General Colin Powell summarised these obligations with the words 'If you break it, you own it.'

Early Days

Nonetheless, the occupation of Southern Iraq began with something of a honeymoon period. When I arrived in Basra in mid 2003, six months or so after the invasion, some local people were still clearly pleased to see us: early in my tour some children literally threw roses. By the end of it, they were throwing stones. Their elder brothers were setting bombs and firing mortars. It got worse.

The Defeat

Over the next years, despite pleas from the local population for the British to protect them, and comply with their legal and moral obligations, ethnic and religious purges cleared Basra of its non-Shi'a population. A bewildered and overwhelmed British Army failed to comprehend what was happening. Militias began to take control of all the important institutions and, as casualties began to mount, the British retreated to their huge base at Basra Airport and ceded control, by agreement, to the mainly Iranian-controlled militia.

In due course a joint Iraqi–US force retook Basra in 'Operation Charge of the Knights' in January 2008, largely without British involvement. A year later, the British were out of Iraq. The US Army took control of Basra in 2009.

One senior British officer described this fiasco to me as 'the biggest kick in the teeth in 34 years of service'. Few would disagree with a British general's assessment that 'We were defeated pure and simple.'[3] The commanding general during the final baleful six-month rotation, General Andy Salmon, was asked by Sir Roderick Lyne at the Iraq Inquiry whether he could identify a strategy. 'We were working to a set of objectives', he said; 'There was no strategic plan that I saw.'[4] Colonel Tim Collins was rather less diplomatic; Iraq, he said, was a 'perfect example of bumbling military incompetence'.[5]

On to Helmand

In 2008, with matters seriously awry in Iraq, the then-head of the British Army, General Sir Richard Dannatt said

> There is recognition that our national and military reputation and credibility, unfairly or not, have been called into question at several levels in the eyes of our most important ally as a result of some aspects of the Iraq campaign. Taking steps to restore this credibility will be pivotal ... and Afghanistan provides an opportunity.[6]

It is fair to say that, at the time of discussion of deeper British military involvement in Afghanistan, the Army was threatened with severe cuts.

---

[3] Unnamed British General quoted in BBC report; Paul Wood, 'Uncertainty follows Basra exit', 15 December 2007, http://news.bbc.co.uk/1/hi/world/middle_east/7145597.stm.

[4] Major General Andrew Salmon's evidence to the Iraq Inquiry, 20 July 2010.

[5] Interview with John Harris, The Guardian, 21 November 2011, www.theguardian.com/uk/video/2011/nov/21/tim-collins-iraq-john-harris [accessed 6 August 2018].

[6] General Sir Richard Dannatt, 'Perspectives on the Nature of Future Conflict', Lecture to Chatham House, 15 May 2009.

Sir Sherard Cowper Coles states that he was told that one reason why the Army were so keen on a large-scale involvement was that 'battalions are coming free from Basra, and we must use them or lose them'.[7] This stress on institutional reputation, or indeed the size and structure of the armed forces, especially the Army, without commensurate examination of national interest comes perilously close to one definition of militarism. It also echoes strongly the selection of the largest 'package' to assist the USA in the invasion of Iraq – with the attendant disastrous consequences.

The 'New Great Game'

The British had been in Afghanistan since the beginning of the war, in 2001. The UK has a strange, long-standing relationship with Afghanistan such that there was a real sense, even when I was there in 2007, that this was the latest round in some kind of spiffing contest.[8] As if to confirm this in Afghan eyes, the first thing the British did was name their main camp in Kabul Camp Souter, after an imperial hero, and regularly conduct ceremonies at the military cemetery, 'home' to those from their first defeat of 1842. It is fair to say that none of this played well amongst Afghans with any sense of history (which is to say most of them).

Deployment to Helmand

In 2006, the UK sent its most aggressive military brigade, 16 Air Assault, to Helmand. When it arrived, Helmand, although run by an extremely corrupt governor, was relatively peaceful. Helmand was also relatively prosperous, being at the heart of the country's extensive drugs-production industry, responsible for nearly half the world's opium for heroin. President Karzai had control of the province, or at least one of his tribal allies did, girls were in school, and it was home to 160 foreign

---

[7] Written evidence of Sir Sherard Cowper Coles to the House of Commons Select Committee on Foreign Affairs, 23 December 2010. See the full report of the Inquiry on the UK's Foreign Policy Approach to Afghanistan and Pakistan, www.publications.parliament.uk/pa/cm201011/cmselect/cmfaff/514/51414.htm [accessed 6 August 2018].

[8] I was not alone. See David Loyn, 'A new Great Game: observations on Afghanistan', *New Statesman*, 13 December 2007, www.newstatesman.com/world-affairs/2007/12/afghanistan-british-taliban [accessed 6 August 2018].

troops,[9] all Americans who had taken no casualties. One officer who had visited Helmand on a reconnaissance told me that his advice in 2005 was 'There is no insurgency there now, but if you want one, you can have one.'[10]

'The Maiwand Thing'

The current president of Afghanistan, Dr Ashraf Ghani, said that, 'If there's one country that should not be in Southern Afghanistan, it is the United Kingdom.'[11] The reason for this was straightforward: the British had a deeply unfortunate history in Southern Afghanistan. Whilst for many of them, it may have been the fourth round of a swashbuckling adventure, for Afghans this was the return of an army that had made a habit of going through the land with fire, rapine, and the sword. In 1880, the Afghans had defeated them at the Battle of Maiwand, a battle which for Afghans has all the resonance of the Battles of Agincourt and Trafalgar, and the Battle of Britain rolled into one.

When this issue was tentatively raised prior to the Helmand move, by one of the Army's more perceptive and informed commanders, Brigadier Andrew Kennett, he was told 'Plans are plans, and there will be no backtracking.'[12]

The Charge Up the Valley

Despite hopes that 'not a shot might be fired', they got their insurgency. An uninformed, undermanned British force, with a badly confused command structure, seriously overextended itself, with small units being sent to remote towns. These immediately came under attack from the Taliban, narco-gangs, and militias, including those allied to the most recent governor of the province, who had been dismissed at Britain's request. Once

---

[9] Wikileaks Report of US Diplomatic conversation with Karzai, 8 September 2007, https://wikileaks.org/plusd/cables/07KABUL2998_a.html [accessed 6 August 2018].

[10] Interview with a senior Special Forces officer, October 2010.

[11] Ashraf Ghani, quoted in Jack Fairweather, *The Good War: Why We Couldn't Win the War or the Peace in Afghanistan* (London: Vintage, 2015), p. 151.

[12] Fairweather, *Ibid.*

again, the Army found itself at sea in a deeply perplexing environment. One leading historian of the campaign, Jack Fairweather, summarised the effects of the first few months on just three towns:

> Sangin's once thriving bazaar was now a pile of rubble ... Musa Qala had also taken a pounding with refugees living in tents in the craters where their homes had once stood. In the North West the town of Now Zad had been entirely abandoned. British fortifications in each town had also grown with outposts surrounded by heaped sacks of earth 15 feet high and topped with barbed wire.[13]

Reports from the Front

A constant stream of upbeat reports gave the lie to such naysaying. In 2007 General Richards said that 'We are winning the fight against the Taliban.'[14] This view was not shared by his American successor as commander NATO forces in Afghanistan, General Dan McNeill, who said that year that 'The British have made a mess of things in Helmand, their tactics were wrong ...'[15] In 2008, the commander of British combat forces in Helmand, Brigadier Mark Carleton Smith, staved off the cynics thus: 'I can therefore judge the insurgency a failure at the moment ... we have reached a tipping point.'[16] US officials were not so confident, reporting in 2008 that 'We and Karzai agree that the British are not up to the challenge of securing Helmand.'[17]

In 2009, as in Basra, the USA came to the rescue. General Stanley McChrystal was keen to know why the US Marines of President

---

[13] Fairweather, *Ibid.*

[14] General David Richards, interview in Fairweather, *Ibid.*

[15] Quoted in Jon Boone, Jonathan Steele, and Richard Norton-Taylor, 'Wikileaks cables expose Afghan contempt for British military', *The Guardian*, 2 November 2010, www.theguardian.com/uk/2010/dec/02/wikileaks-cables-afghan-british-military [accessed 6 August 2018].

[16] Mark Carleton-Smith, interview with Thomas Harding, 'Afghan insurgents "on brink of defeat"', *The Telegraph*, 1 June 2008, www.telegraph.co.uk/news/newstopics/onthefrontline/2062440/Afghanistans-Taliban-insurgents-on-brink-of-defeat.html [accessed 6 August 2018].

[17] Senior US State Department official quoted in Jon Boone, Jonathan Steele, and Richard Norton-Taylor, 'Wikileaks cables expose Afghan contempt for British military', *The Guardian*, 2 November 2010, www.theguardian.com/uk/2010/dec/02/wikileaks-cables-afghan-british-military [accessed 6 August 2018].

Obama's 'surge' were being sent to Helmand.[18] The reason, he found, was that, as he put it himself, the British were 'essentially besieged inside sandbagged outposts the Taliban surrounded'.[19]

From a relatively peaceful, and by Afghan standards prosperous, province in 2006, Helmand was by 2009 by far the most intense combat zone on Earth, and one in which the British were losing, and losing badly. In one sense, 2009 marked the end of the *British* battle for Helmand. They had failed. Now most of the foreign force was composed of US Marines.

This did not stop General Richards from declaring that 2010 would be another 'turning point'.[20] When that failed to materialise, he promised that 2011 contained yet another 'tipping point'. The campaign was 'on track', and the Taliban 'were completely unhinged'.[21] 'The planets were aligned', said Brigadier Sanders in 2012, 'We are on the cusp of delivering durable success in Central Helmand.'[22]

David Cameron declared 'mission accomplished'[23] in December 2013, and a year later the British flew out of their vast base in the desert, Camp Bastion, for the last time. By that time, the central idea was not to control territory; it was rather to provide women's education, 'extend the remit

---

[18] Rajiv Chandrasekaran, *Little America: The War within the War for Afghanistan* (London: Bloomsbury, 2012). A relevant excerpt can be read in *The Washington Post*, 22 June 2012, www.washingtonpost.com/world/war-zones/little-america-excerpt-obamas-troop-increase-for-afghan-war-was-misdirected/2012/06/22/gJQAYHrAvV_story.html [accessed 6 August 2018].

[19] General Stanley McChrystal interview with Peter Foster, 'US General says Britain risks "special relationship" if it cuts military', *The Telegraph*, 17 January 2013, www.telegraph.co.uk/news/worldnews/northamerica/usa/9808791/US-general-says-Britain-risks-special-relationship-if-it-cuts-military.html [accessed 6 August 2018].

[20] General David Richards, interview with Con Coughlin, 'General Sir David Richards: forces reach "turning point" in Afghanistan', *The Telegraph*, 26 February 2010, www.telegraph.co.uk/news/worldnews/asia/afghanistan/7326145/General-Sir-David-Richards-Forces-reach-turning-point-in-Afghanistan.html [accessed 6 August 2018].

[21] UK Ministry of Defence statement, 'Home from Helmand, UK Commanders Review HERRICK 14', 19 October 2011, www.gov.uk/government/news/home-from-helmand-uk-commanders-review-herrick-14 [accessed 6 August 2018].

[22] MOD Announcement, 'On the Cusp of Durable Success in Helmand', 8 June 2012, www.gov.uk/government/news/on-the-cusp-of-durable-success-in-helmand [accessed 6 August 2018].

[23] Rowena Mason, 'David Cameron declares "mission accomplished" in Afghanistan', *The Guardian*, 16 December 2013, www.theguardian.com/uk-news/2013/dec/16/afghanistan-mission-accomplished-david-cameron [accessed 6 August 2018].

of GIROA',[24] or develop courts. It was to provide basic capability for the Afghan Army and security forces.

Helmandi Aftermath

The war was over for us, at least for the time being. But for Helmandis and Iraqis it continues even today.[25] As for the Taliban (known as the 'Aslee' or real Taliban to distinguish them from the majority of 'part-time' local insurgents), and other assorted groups and tribes fighting in Helmand, none were party to the various 'tipping' and 'turning' points and regular announcements of their imminent defeat.

When the British left, the Taliban and various other militias hostile to the Afghan government controlled about 70% of the province. Today this figure is closer to 90%. Needless to say, it is too dangerous for Western, or, indeed, impartial Afghan, journalists to go there. President Karzai, not always the most measured of commentators, said in 2014 that the British had created a 'vast countryside of deprivation and anger'.[26] In the words of one US Special Forces officer, 'The British wrote a cheque they couldn't cash.'[27]

Echoing the comments of General Salmon concerning Iraq, Brigadier Butler, the commander of the first deployment to Helmand, said 'We were never really clear what the strategic objectives actually were.'[28] If the commander did not know, how could ordinary soldiers have

---

[24] GIROA – the common acronym for the Government of the Islamic Republic of Afghanistan.
[25] The final version of the strategy involved training the Afghan Army to fight the war. However, 97% of the Army, the only credible security institution in Afghanistan, are not Southern Pashtun, and they are officered almost entirely by Tajiks; they are as foreign to Helmandis as the British were. Without the combat support, medical, air cover, and command and control arrangements of a modern army, which were all absent, they are little more than groups of brave men with rifles; much like their opponents.
[26] President Karzai, interview in *The Sunday Times*, 2 February 2014. Full transcript, http://afghanistanembassy.org.uk/english/full-transcript-of-president-karzais-interview-with-british-newspaper-the-sunday-times/ [accessed 6 August 2018].
[27] Private communication with the author, August 2014.
[28] Operation Herrick Campaign Study; reported in *The Times*, 16 January 2016, following a Freedom of Information request. The original quotation is to be found on p. 1-1-3 in the report, www.gov.uk/government/uploads/system/uploads/attachment_data/file/492757/20160107115638.pdf [accessed 6 August 2018].

any idea? Let me now turn to those who were responsible for those objectives.

Generalship, Strategy, and Politics

US General Daniel Bolger summarised the case for the prosecution as follows:

> Our primary failing in the war involved generalship. We ...
> demonstrated poor strategic and operational leadership ... some might
> blame the elected and appointed civilian leaders. There's enough fault
> to go around. But I know better, and so do the rest of the generals. We
> have been trained and educated all our lives in how to fight and win.
> This was our war to lose, and we did.[29]

The case expounded here is simple and straightforward: the nature, scale, and conduct of both campaigns were decided by senior British military officers, not politicians. These were military campaigns, run by military officers.

Now, most senior officers are decent and intelligent men (they are almost all men). They vary in quality. To quote one of their number, General Graeme Lamb – who achieved a good deal of success in Iraq as part of General Petraeus' team – 'The majority I would rate as fair, a few I would gladly join and assault Hell's gate, and some I wouldn't follow to the latrine.'[30]

You may recall that General Salmon, one of the better generals of the Iraq War, said that 'There was no comprehensive strategic plan as far as I could see'[31] right at the end of Britain's Iraq War. The first commander in Helmand, Brigadier Ed Butler, has said that 'We were never really

---

[29] Daniel P. Bolger, *Why We Lost: A General's Inside Account of the Iraq and Afghanistan Wars* (Dublin: Houghton Mifflin Harcourt, 2014), p. xv.

[30] Graeme Lamb, 'On Generals and Generalship', in Jonathan Bailey, Richard Iron, and Hew Strachan (eds.), *British Generals in Blair's Wars* (Aldershot: Ashgate, 2013), pp. 143–157 on p. 147. Lamb is referring specifically to his fellow contributors to that book. The publication history of *British Generals in Blair's Wars* is instructive. Originally it had twenty-six contributors. The Ministry of Defence refused to clear the contributions of seven who were still serving. Consequently their essays were not included.

[31] Major General Andrew Salmon, evidence to the Iraq Inquiry, 20 July 2010.

clear what the strategic objectives actually were, and how these might be translated into resourced tactical actions on the ground.'[32] As the great General Clausewitz put it: no one starts a war, or rather no one in his senses ought to do so, without being clear in his mind what he intends to achieve by that war and how he intends to conduct it.

You will be relieved to know that I do not intend to dwell on strategy. Suffice it to say that you need one. However, strategy should be the province of political guidance, guided and supported by competent and realistic military advice. I will illustrate this below.

The Political–Military Link

Prior to his deployment as High Commissioner of Malaya in 1952 during the Malayan Emergency, General Gerald Templer insisted on political guidance as to the strategy at the heart of the war. He wrote to Oliver Lyttleton, the Foreign and Colonial Secretary, as follows:

> The general object is obvious – viz to restore law and order and to bring back peace to the Malay Federation. In order that this object may be achieved, I am clear as to what must happen from the military point of view. I am not at all clear as to what [HM Government] IS aiming at from the political point of view ... I must have a clear policy to work on.[33]

He received a detailed and precise reply in the terms he had requested. There is no record of any senior British commander requesting, indeed insisting upon, similar guidance as to the political objective. It may well be that there was no political strategy, as indeed seems the case. That is a separate point. It was the duty of senior commanders to ensure clarity, and if such clarity was absent then insist, as Templer did, that such clarity was produced.

The point here is not that the Generals, Admirals, and Air Marshals – none of whom are trained or experienced in politics or grand strategy – should have come up with policy. Not at all. What they should have done, but

---

[32] Brigadier Ed Butler, quoted in Deborah Haynes, 'Arrogant Army chiefs blamed for "messy" campaign in Helmand', *The Times*, 18 January 2016.
[33] Minute by Sir G. Templer to Mr Churchill PREM 11/639 from *British Documents on the End of Empire* (London: HMSO, 1995), p. 372.

didn't, was to insist that there was clear policy and not some set of vague aspirations. In the absence of that, it would have been their duty to advise that matters were unclear. They were not in a position to design a workable campaign plan in the absence of clear policy, and the armed forces could not guarantee success.

'The Chiefs of Staff View Was That They Could Do It ...'

What happened instead? Sir Kevin Tebbit was Permanent Under-Secretary at the Ministry of Defence, and gave evidence to the Iraq inquiry concerning the idea of a large deployment to Afghanistan. Whilst an equally large deployment to Iraq was already in place, in combat and in trouble in 2005,

> ... their view was they could do it and were satisfied that they could manage that deployment within the resource[s available]. We had a meeting with the Chiefs of Staff. I was concerned. The weight of the views of the Chiefs of Staff themselves was in favour. Since it was they who would have to ensure they could do this, I did not press my concerns further.

Tony Blair in his evidence was succinct:

> ... they were up for doing it and they preferred being right in the centre of things.

It is the duty of senior commanders to advise in a sober and realistic fashion as to the necessity, inter alia, for a realistic strategy and then to advise, in a similar fashion within the pertaining constraints, which in this case were considerable, whether that strategy was in fact achievable. Those commanders failed in that duty. It is a fundamental failure.

There are multiple other examples of Chiefs of Staff informing politicians as to risks and constraints. Field Marshal Lord Alanbrooke was constantly ensuring that Churchill's plans remained within the realms of the possible, and that the risks and constraints were fully explained. His catchphrase was 'I flatly disagree.' His biographer, David Fraser, says that this great man had a 'skeptical attitude to strategic planning assumptions' derived from his experiences in the First World War.

Institutional Dishonesty

Prior to and during the Falklands War, all the Chiefs of Staff were explicit in their advice to the Prime Minster that the operation to recover the Falklands would be a high-risk venture, and might not be possible; papers set out for ministers what could go wrong, the possible high casualties, and what needed to be done to forestall, as far as possible, potential problems.[34] We have seen how senior officers, in their desire to be seen as at the 'heart of things', were reluctant to give negative advice, and indeed recommended large-scale deployments where common sense and mathematics might indicate a modest view was appropriate.

This was not happening within the higher reaches of the MOD; Sir Sherard Cowper Coles, who as Ambassador to Afghanistan during this period would have had a clear insight into the inner workings of government, said this:

> I saw in my three and a half years papers that went to Ministers that were misleadingly optimistic. Officials and Minsters who questioned them were accused of being defeatist or disloyal in some way.[35]

At the lower, tactical levels there was a similar aversion to being told bad news. The views of a soldier friend of mine might be worth considering: 'In my reports I'd say that our GDA was achieving [nothing] ... by the time it came out at Brigade, we were pressing on, progress was being made and all would be well.'[36]

Accountability Is Important

In the law or medicine the delivery of unrealistic advice results in adverse consequences both for those requesting the advice and for those delivering it. After all, while the price of failure in finance might be the collapse

[34] See for example the contributions of Admiral Leach and Field Marshal Bramall to 'The Falklands War', CCBH Oral History Programme symposium held at the Joint Services Command and Staff College (Symposium Paper published by the Centre for Contemporary British History 2005).

[35] Evidence of Sir Sherard Cowper Coles to House of Commons Select Committee on Foreign Affairs, 9 November 2010. See the report of the Inquiry on the UK's Foreign Policy Approach to Afghanistan and Pakistan, www.publications.parliament.uk/pa/cm201011/cmselect/cmfaff/514/51414.htm [accessed 6 August 2018] at Q99.

[36] Interview with infantry officer, June 2014.

of a bank, military failure has rather wider consequences. This is why accountability is important. If we do not know who is accountable, how do we know who is in charge. If we do not know who is in charge, whom do we look to for guidance?

Soldiers are told in basic training that it is vital for an officer to take responsibility, 'to have broad shoulders'. Culpability is irrelevant; responsibility is what counts. In the Iraq inquiry, no fewer than thirty-eight senior generals or equivalent gave evidence. Yet, in an ever-shifting environment of changes of command and unclear lines of authority, who was in charge? Who was accountable?

Are Numbers Important?

It is not in our military tradition for accountability to be evaded in such fashion. Do numbers have an impact? We have six times more General officers (pro rata) than the USMC, we have three times more General officers than the US Army (pro rata), and ten times more than the Israeli Defence Forces.

From Georgian days, when Admiral Byng was shot by a firing squad for failing to take Minorca in 1756, all the way up to the Falklands and beyond, matters were arranged so that it was amply clear who was responsible, and accountable for victory, or defeat. This accountability, in the event of failure, took the form of removal from post. During the Second World War, the remarkable Chief of the Imperial General Staff Alan Brooke regularly dismissed senior officers who had failed. Brooke said that 'Half of our Corps and Divisional Commanders are totally unfit for their duties. And yet if I were to sack them I would find no better.'[37] A rather more recent senior officer, now retired, was rather more blunt: 'The level of incompetence is so great and has been occurring for so many years now, that it's systemic and it's become cultural.'[38]

The Difference between Accountability and Culpability

There is a tradition in the Royal Navy that, whoever is at fault, it is the Captain of a ship who must take the responsibility, usually in the form of

[37] Alan Brooke (ed. Alex Danchev), *War Diaries 1939–1945* (London: Weidenfeld and Nicholson, 2002), entry for 31 March 1942.
[38] Major General Andrew Mackay, 12 January 2012.

Court Martial. One example of this was the grounding of HMS *Astute* in 2008. Even though the Captain was not in the control room, he bore ultimate responsibility.

The Camp Bastion Raid

In 2011, Camp Bastion was raided very successfully by a Taliban *commando* team who destroyed six jet fighters and seriously damaged many more aircraft, also killing two US servicemen, including the American squadron commander. Perimeter security was the *responsibility* of British forces. As a result of the ensuing, damning US investigation, the two most senior generals in Southern Afghanistan were not simply dismissed from their posts, but made to resign their commissions.

Ironically, the British senior officer responsible for base security was promoted. In fairness to him, he was not culpable for the raid; culpability lay further down the chain of command, as it did for the US generals. But this is unimportant. He was *responsible*, and traditional military or, indeed, business practice would indicate, also *accountable*. In a decade of military failure and disaster, no senior officer has been held to account; none was ever dismissed; none resigned. In any other society, this would be termed a 'culture of impunity'. Indeed, given that every general found a mention in that rather curious British institution the 'Honours List' at the end of his (and it is always 'his') tour of duty, one might almost describe it as an 'all must have prizes' culture. In addition to the inevitable honours, no fewer than half of all British generals commanding UK forces in the fiasco that was Iraq were promoted.

In a society such as ours with so little questioning of military matters, or veiling of criticism in the shrouds of 'our boys' and 'respect', it is possible to evade close scrutiny. In other countries, however, where soldiers are seen in a rather more realistic light, this is not the case.

The Winograd Inquiry

In Israel, for example, defence is a matter of vital concern. Soldiers are not 'the other', reified as heroes; each Jewish citizen is a soldier and few have any illusions about senior officers, or indeed have much respect for notions of immunity of the institution from criticism.

If there is failure or defeat, as was clear, for example, after the Lebanon War of 2006, there has to be accountability, quickly applied, in the national interest. There was a 'flawed performance by the army, deficient preparedness', and, 'because of the conduct of the high command, it failed to provide an effective military response'.[39] As a result, the Chief of Staff was dismissed, other senior officers removed, and the Prime Minster was forced to resign. The report appeared a full two years after the end of the war. Many observers in Israel thought that this was far too long.

Intelligence

The Chinese strategist Sun Tzu famously wrote 'Know the enemy, know yourself and victory is never in doubt, not in a hundred battles. Know neither yourself nor your enemy, and you will lose every battle.' Intelligence is traditionally seen by soldiers in a rather particular way. At its heart, the objective is to 'find, fix, and finish the enemy'. In the kind of war for which the British military is structured, this fits well. Sadly, the British Army, Navy, and Air Force are structured and equipped to fight wars against armies, navies, and air forces much like themselves: large and highly technically capable, but with a high degree of intellectual and conceptual inertia.

Know Thyself …

The UK military is excellent at observing, deciding, and acting. Orientation, however, is a decided weak point, and intelligence is all too often seen not as supporting decision-making, but, as one senior officer put it to me recently, supporting the decision once it has been made![40]

At the lower tactical level in Iraq, our vast intelligence apparatus was exclusively focused on secret information, often from highly unreliable but very 'secret and therefore privileged sources', concerning supposed enemies within Sunni resistance groups. I spent many long evenings on

---

[39] Israel Ministry of Foreign Affairs, 'Winograd Committee Submits Final Report', 30 January 2008, www.mfa.gov.il/mfa/mfa-archive/2008/pages/winograd%20committee%20submits%20final%20report%2030-jan-2008.aspx [accessed 6 August 2018], paragraph 20.

[40] Interview with senior serving officer, December 2015.

raids looking for such people, most of whom did not exist. When the real enemies, Shi'a death squads, hit us, it was a rude awakening indeed.

Had we spent time listening to the people of Basra, who lived in terror not of the enemy we wanted to fight, but of brutal fundamentalist enforcers, the situation in Basra might have been very different. This would have required equal stress to be placed upon so-called 'open source intelligence', known to all as covert secret intelligence.

To do so, however, would have required a fundamental change in intelligence culture, where the objective was not to kill enemies or achieve short-term tactical objectives but understand who and what we were (which was not, as I heard one general say, 'the biggest and best gang in Basra'). Rather more recently, Michael Martin described how the British in Helmand were seen by many local actors as 'the biggest, best-armed, richest and stupidest gang in the province'.[41] As he points out in *An Intimate War*, this lack of understanding lent itself to the British being treated as 'useful idiots' by some of those actors with scores to settle or causes to advance. What better way of being rid of a rival or enemy than to report him to the occupying forces as a 'Taliban' operator? This highly destructive and damaging factor was exacerbated by a constant demand, particularly on the part of Special Forces, for targets to be added to the 'Joint Priority Effects List', or JPEL.[42] This was essentially a list of those to be killed or captured.[43]

Know the Enemy ...

Knowing precisely who our enemies were was difficult. In the early battles of the campaign – and indeed throughout – it was thought that the British were fighting the Taliban. We now know that most of those who were fighting us were not those we know as Taliban at all. In Sangin, at the time when the town was destroyed following British entry there, it was thought that those fiercely attacking the beleaguered British base

---

[41] Interview with Michael Martin, February 2016.
[42] Interview with UK military officer, March 2016.
[43] For a good description of how the JPEL operated, see *Der Spiegel* (Online International Edition), 28 December 2014, www.spiegel.de/international/world/secret-docs-reveal-dubious-details-of-targeted-killings-in-afghanistan-a-1010358.html [accessed 6 August 2018].

were Taliban, whereas we now know that they were local tribal groups, incensed at our perceived support of a well-known narco-criminal who ran the town on behalf of the 'government', *not* the Taliban.

Know yourself, says the man, know your enemy. Both of these require a fundamental shift in the approach to intelligence which remains very much 'enemy-focused'. A close friend of mine, an Arabic-speaking scholar and soldier who was seriously injured in Iraq, says that in environments such as Basra, or Helmand, we are playing what he calls games of three-dimensional chess. To do so successfully requires systems to accommodate that human complexity. It also needs the right people, or the right players if you will. Even if such people exist today, all too often they are not used. Indeed, they are often ignored or actively sidelined, or worse, for delivering unwelcome views.

Playing this three-dimensional game also requires a change in mentality. General Sir Rupert Smith, himself a very successful general, has called this 'opening networks'.

Opening Networks: Building a 'Challenge Culture'

We have a fine cadre of middle-ranking officers coming through our system. The challenge will be to retain them, and not allow ossified one-dimensional promotion systems that encourage conformity and intellectual inbreeding to block the progress of what will be the most experienced, critically minded generation of officers we have seen since the Second World War. The signs are not particularly promising; no fewer than eighteen of the last nineteen full generals appointed since 1990 have been public schoolboys selected from five regiments.[44]

'I want to hear solutions', as my old commanding officer used to tell me. I'd like to look briefly at what might be done to open the necessary networks and build what one general has called 'the post-modern warrior'.[45] In a world of ever greater complexity, where what is required is

[44] There are about fifty regiments in the army, depending on how the term 'regiment' is interpreted. Some, such as the 'Royal Regiment of Artillery' are huge and themselves comprise many smaller regiments. Such is the nature of an organically grown institution such as the British Army.

[45] General Sir John Kiszely, *Post-Modern Challenges for Modern Warriors*, Shrivenham Papers No. 5, September 2007.

acute awareness, those closed and exclusive networks need to be opened, and critically broader approaches must be taken.

Few would contend that, at the higher levels of command and policy, the idea of a challenge is part of the cultural backdrop to Whitehall in general, or the Ministry of Defence in particular. It was perhaps revealing that Sir Kim Daroch, the National Security Advisor, stated that he did not know what a 'red team' was.[46] It could be, of course, that he was simply not familiar with the neologism, rather than the concept.

One recently retired senior infantry officer agreed that 'We still have a culture of operational honesty, but we are not afforded the same level of honesty coming down to us. There was a real feeling that we were being messaged, that it was patronising and inappropriate and did not give soldiers the credit for their self-evident intelligence and street-smarts.'[47] There seems to be a feeling amongst those involved in defence that operational honesty is for those towards the bottom of the ladder and not those at the top. Much like accountability.

Educating Our People Better

One way to begin to address both the question of challenge and the related issue of the 'problem of knowledge' would be to educate our senior soldiers, sailors, and airmen better. Education is different from training. No British soldier is short of teaching in necessary military skills. Training teaches you how better to cope with things you know. Education teaches you to be ready to deal with things you don't. That way, a surprise need not be a shock.

The kind of education I advocate does not mean in-house MAs at military colleges, institutions rather intellectually akin to conservative theological academies. One way of educating our officers is to pay for the very best of them to pursue three years to PhDs. Done properly, a five-year

---

[46] 'I'm sorry, I don't know what a "red team" is'; Seventh Report of Commons Select Committee on Defence (Session 2013-24), 'Towards the Next Security and Defence Review: Part 1', evidence of Sir Kim Daroch at Ev Q140, www.publications.parliament.uk/pa/cm201314/cmselect/cmdfence/197/19702.htm [accessed 6 August 2018].

[47] Interview with the author, July 2015.

programme for 100 officers per year, like the American Advanced Civil Schooling programme, would cost less than a single fighter aircraft.

Making an argument for the importance of a graduate – or, indeed, any good – education to an academic audience is pushing at an open door. General David Petraeus, like hundreds of other officers in the US military, benefited from the Advanced *Civilian* Education System programme. In an article titled 'Beyond the cloister' he says that he was

> ... exposed to diverse and sometimes hostile views, and sometimes diverse cultures; just as the best way by far to learn a foreign language is to live in the culture where the language is spoken, so the best way to learn about other worldviews is to go and live in another world; grad school forces a person to redefine upward one's own internal standards of excellence.[48]

Another very successful senior officer, General H. R. McMaster, did a military-funded PhD at the University of North Carolina at Chapel Hill. He gave a talk to students there, saying that his time at Chapel Hill had prepared him for combat in Iraq: 'It was here that I think I learned to ask the right questions ... about complex issues.'[49] His tour in Tal Afar as commander of the 3rd Armoured Cavalry Regiment stands as one of the very few success stories of the War in Iraq. (Incidentally, his PhD was later turned into a book, *Dereliction of Duty: Lyndon Johnson, Robert McNamara, The Joint Chiefs of Staff, and the Lies That Led to Vietnam*. Not surprisingly, General McMaster has a rare reputation of challenging the status quo.)

The occasional British officer is sponsored through a graduate programme. Captain Michael Martin, a Pashto-speaking multi-tour veteran of Helmand, wrote a PhD on Helmand's tribes and history. It was critical of the British Army, asserting what had been obvious to anyone involved in operations, namely that the British lack of understanding of

[48] David H. Petraeus, 2007, Beyond the cloister, *The American Interest* 2(6), www.the-american-interest.com/2007/07/01/beyond-the-cloister/ [accessed 6 August 2018].

[49] Thomas Ricks, 'BG H. R. McMaster 'How being a Grad student at Chapel Hill prepared me for combat', *Foreign Policy*, 12 April 2010, http://foreignpolicy.com/2010/04/12/bg-h-r-mcmaster-being-a-grad-student-at-chapel-hill-prepared-me-for-combat/ [accessed 6 August 2018].

the environment was such that the army had often been used as 'useful idiots' by groups seeking to gain advantage over rivals.

Captain Michael Martin was forced to resign his commission in 2014. The British Army took the somewhat bizarre view that his book,[50] commissioned by them as a PhD project and now the leading work on Helmand history and society, undermined their necessary (if entirely false) narrative of relentless success in the Province.[51] As a highly experienced Pashto- and Burmese-speaking officer, Dr Martin was a real loss to the UK defence establishment, who are not over-equipped with linguistically talented, field-experienced, and academically minded, constructively critical officers.

Sadly, what happened to Michael Martin says much of our armed forces' approach to 'open networks'. This needs to change, and change soon. We are not short of security challenges, and are not preparing our leaders to deal with them, so that, whilst there may always be surprise, there won't be shock. Whilst there are endless training courses available to senior officers, there is currently little evidence that there is a clear understanding of the importance of education with its concomitant critical thinking within the higher reaches of the UK's armed forces.

Whether the UK decides to take education seriously remains to be seen. However, there is plenty of evidence that peers and potential enemies see the importance of *understanding* as a necessary complement to capability. The USA has an extensive programme to fund military officers to do graduate work. Indeed, there are more US military officers doing PhDs funded by their Defense Department in UK universities than British ones. Another figure that should give pause for thought is that there are more Chinese officers at US graduate schools than American officers.[52]

[50] Michael Martin, *An Intimate War: An Oral History of the Helmand Conflict* (London: Hurst, 2014).
[51] See Tom Coghlan, 'MOD tries to block its own book on Helmand', *The Times*, 9 April 2014, www.thetimes.co.uk/tto/news/uk/defence/article4058238.ece (paywall) [accessed 6 August 2018].
[52] United States Joint Forces Command, *Joint Operating Environment* (Norfolk, VA: United States Joint Forces Command, 2008), p. 27.

Conclusion

The opponents we may face are formidable: from Russia's hybrid war – or the kind of conflict we see in Ukraine – to the complex and highly capable threat posed by China's so-called 'three warfares', with cyber warfare as a leading element. To the extent that involving ourselves in fighting in Iraq and Syria is in our national interest, it is highly likely to require the kind of skills and approach we so signally lacked in Iraq and Afghanistan. We will need people who are equipped to advise and lead and succeed in highly complex contests taking place in arenas of which we have little or no experience.

In the very different, knife-edge world of the Cold War, Professor Michael Howard said this:

> If there were to be another war, the first battle may be the last ... the social changes of our time may so transform the whole nature of warfare that the mode of thought of the military professional today will be at best, inadequate or, at worst, irrelevant. This is the kind of change for which we must be prepared and able, if necessary to adjust ... the alternative is disappearance and defeat.[53]

In the ever-more challenging arenas facing us, we need people who can cope with swift change and multi-dimensional threats, people for whom surprise does not turn into shock. We've a unique opportunity now to take advantage of an excellent cadre of experienced and intelligent officers. The challenge for their leaders is first to ensure that the best are retained, for very many of them have left the armed forces or plan to do so soon. Equally important is to ensure that those who do remain are properly prepared for the complex challenges they will face.

---

[53] Michael Howard, 1974, Military science in an age of peace, *RUSI Journal* 119(1), 7–8.

6 Games for the Brain

BARBARA J. SAHAKIAN, LAURE-SOPHIE CAMILLA
D'ANGELO, AND GEORGE SAVULICH

Introduction

Mental ill health is the single largest cause of disability in the UK.[1] In addition to causing distress to individuals and their families, mental health problems are an enormous financial burden to society and the economy. The wider economic costs of mental illness in England are estimated at £105.2 billion each year, with three disorders, depression, dementia, and schizophrenia, costing approximately £23.8 billion, £17 billion, and £13.3 billion in total annual costs, respectively. This surpasses both cancer and cardiovascular disease.[2] Thus, there is a clear need for new strategies to effectively prevent, diagnose, and treat mental health disorders.

Paradigm Shift: Early Detection and Early Effective Treatment

In order to reduce the burden of mental ill health, both to the individual and to society, neuroscience and mental health policy has recently emphasised the need for a 'paradigm shift' in how we detect and treat mental health disorders.[3] It is increasingly acknowledged that, as mental health disorders become chronic and relapsing, they more severely

---

[1] N. A. Fineberg, P. M. Haddad, L. Carpenter, B., Gannon, R., Sharpe, A. H. Young, and B. J. Sahakian, 2013, The size, burden and cost of disorders of the brain in the UK, *Journal of Psychopharmacology (Oxford, England)* 27(9), 761–770.

[2] P. Y. Collins, V. Patel, S. S. Joestl, D. March, T. R. Insel, and A. S. Daar, 2011, Grand challenges in global mental health, *Nature* 475(7354), 27–30.

[3] J. Beddington, C. L. Cooper, J. Field, U. Goswami, F. A. Huppert, R. Jenkins, and S. M. Thomas, 2008, The mental wealth of nations, *Nature* 455(7216), 1057–1060.

impact on the life course and become difficult, if not impossible, to treat. For example, the time it takes until obsessive-compulsive disorder is diagnosed can be as much as 17 years.[4] Similarly, schizophrenia affects young adults and can be lifelong, so early detection is important to ensure the best possible outcome. Depression can also recur throughout life. In order to preserve mental wellbeing and ensure that people have a good life course, mental illnesses should be a particular target for early detection and early effective treatment. Indeed, the Foresight Project on Mental Capital and Wellbeing by the UK Government Office for Science recognised the importance of good brain health to ensure a flourishing society and emphasised the importance of early identification, treatment, and, eventually, prevention of mental health disorders (Figure 6.1).[5] The project further emphasised that biomarkers, including cognitive, neuro-imaging, genetic, cerebrospinal fluid, and blood-based measures, will be critical to achieve this and was recently followed up by the Government Office for Science.[6]

Restoring Cognitive Function

It is now recognised that psychiatric disorders are disorders of cognition, motivation, and their interaction. Cognitive impairments in psychiatric disorders include attentional biases, learning problems, memory impairments, dysfunctional reward systems, and lack of top-down cognitive control by our prefrontal cortex.[7] It is perhaps not surprising that mental

[4] E. Hollander, E. Doernberg, R. Shavitt, R. J. Waterman, N. Soreni, D. J. Veltman, and N. A. Fineberg, 2016, The cost and impact of compulsivity: a research perspective, *European Neuropsychopharmacology* 26(5), 800–809.

[5] Beddington, *et al.*, The mental wealth of nations.

[6] B. J. Sahakian, 'What Next for Mental Health and Wellbeing?', 23 October 2017, https://foresightprojects.blog.gov.uk/2017/10/23/what-next-for-mental-health-and-wellbeing/ [accessed 6 August 2018].

[7] A. B. Brühl and B. J. Sahakian, 2016, Drugs, games, and devices for enhancing cognition: implications for work and society, *Annals of the New York Academy of Sciences* 1369(1), 195–217; B. J. Sahakian, A. B. Bruhl, J. Cook, C. Killikelly, G. Savulich, T. Piercy, *et al.*, 2015, The impact of neuroscience on society: cognitive enhancement in neuropsychiatric disorders and in healthy people, *Philosophical Transactions of the Royal Society B: Biological Sciences* 370(1677), 20140214; G. Savulich, T. Piercy, A. B. Bruhl, C. Fox, J. Suckling, J. B. Rowe, and B. J. Sahakian, 2017, Focusing the neuroscience and societal implications of cognitive enhancers, *Clinical Pharmacology and Therapeutics* 101(2), 170–172.

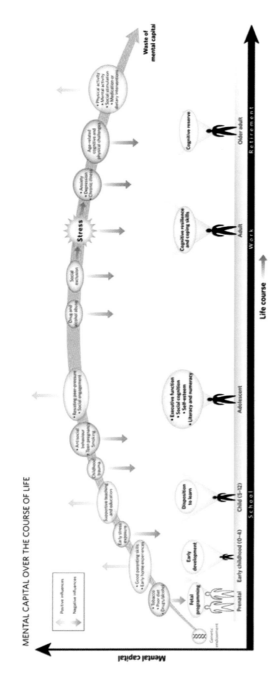

FIGURE 6.1 Mental capital over the course of life. Synthetic view of the mental capital trajectory and factors that may act upon it. Reprinted from Beddington *et al.* (2008) The mental wealth of nations. *Nature 455*, 1057–1060. Copyright (2008), with permission from Springer Nature (www.nature.com/nature/journal/v455/n7216/full/4551057a.html).

disorders feature impairments in top-down cognitive control since we now know that our brains are still in development during late adolescence and early young adulthood.[8] Therefore, environmental stressors will have a particular impact during this neurodevelopmental phase. Indeed, many neuropsychiatric disorders are of neurodevelopmental origin, with an onset or prodromal stage in childhood or adolescence. Mental disorders disproportionately affect the young, with 75% of illnesses having an onset before the age of 24 years.[9] Current treatments for a wide range of mental health disorders rarely target cognitive symptoms, even though these have a significant impact on the functionality and wellbeing of individuals (e.g. not finishing school or attaining only low grades, inability to train for a job, lower academic achievement, absence from work, or early retirement). Therefore, novel treatments that focus on improving cognitive dysfunction would improve the functionality of patients and, in doing so, help to alleviate some of the economic and societal costs associated with disease burden.

Cognitive Enhancement in Neuropsychiatric Disorders

Early Detection

Novel cognitive-enhancing and neuroprotective treatments for psychiatric and neurodegenerative disorders will need to be initiated early, if they are to be successful. Early detection and treatment was emphasised by the former Prime Minister David Cameron in his speech on dementia in 2012, in which he stated that 'Only around 40 per cent of those with dementia know they have it.' He also noted that 'You can help people live independently for longer or even put the brakes on their decline.'[10]

---

[8] N. Gogtay, J. N. Giedd, L. Lusk, K. M. Hayashi, D. Greenstein, A. C. Vaituzis, and P. M. Thompson, 2004, Dynamic mapping of human cortical development during childhood through early adulthood, *Proceedings of the National Academy of Sciences of the USA* 101(21), 8174–8179.

[9] R. C. Kessler, M. Petukhova, N. A. Sampson, A. M. Zaslavsky, and H.-U. Wittchen, 2012, Twelve-month and lifetime prevalence and lifetime morbid risk of anxiety and mood disorders in the United States, *International Journal of Methods in Psychiatric Research* 21(3), 169–184.

[10] Transcript: Prime Minister's speech to the Dementia 2012 conference, www.gov.uk/government/speeches/transcript-prime-ministers-speech-to-the-dementia-2012-conference [accessed 6 August 2018].

Currently, a diagnosis of Alzheimer's disease requires the development of multiple cognitive deficits, including memory impairment, as well as problems with at least one other area of cognition, such as language, motor sequencing, or naming.[11] Furthermore, this cognitive decline has to be sufficiently severe that it causes impairment in occupational and social functioning. This means that, by the time the disease is detected, patients already have significant problems at work and at home. Thus, there is a great need to detect and treat patients earlier in order to prevent this sort of decline in functional outcome. Early treatment would allow patients to stay in their own homes for longer, which is beneficial to the individual. Furthermore, early detection of Alzheimer's disease is cost-effective for the National Health Service (NHS). Early assessment and treatment of Alzheimer's disease saves society £7,741 per patient, of which £3,600 is in direct healthcare costs.[12] Early treatment of dementia is particularly important as our ageing population means we can only expect these costs to increase in future.

Cholinesterase inhibitors, such as donepezil (Aricept), are currently approved for the treatment of cognitive symptoms in Alzheimer's disease. Whilst these drugs improve attention and concentration, they are less effective at treating memory problems. Early on in the course of Alzheimer's disease, cell loss in the hippocampal formation leads to deficits in episodic memory, which is important for everyday functions, such as trying to find your car in a multi-storey car park or trying to remember where you left your keys in the house. Later on in the course of the disease, cell loss in the temporal neocortex results in problems in semantic memory, which is important in knowledge for facts (e.g. that England is part of the United Kingdom). Episodic memory is related to functional outcome in Alzheimer's disease, and so novel and improved symptomatic treatments to address memory problems could have a positive impact on functionality and wellbeing. In addition to drugs that target cognitive

[11] American Psychiatric Association, *Diagnostic and Statistical Manual of Mental Disorders. Fifth Edition. DSM-5* (Washington, D.C.: American Psychiatric Association, 2013).
[12] D. Getsios, S. Blume, K. J. Ishak, G. Maclaine, and L. Hernández, 2012, An economic evaluation of early assessment for Alzheimer's disease in the United Kingdom, *Alzheimer's & Dementia: The Journal of the Alzheimer's Association* 8(1), 22–30.

symptoms, it is hoped that the development of novel neuroprotective drugs will slow or actually halt the underlying disease process. However, in order for these drugs to be most effective, they will need to be given during the early or prodromal stages of Alzheimer's disease (e.g. mild cognitive impairment), so that individuals can be treated while they still maintain a good quality of life and their cognition, wellbeing, and functionality at work and at home are still intact. It is therefore essential to develop innovative tools to detect the disease early to achieve the best clinical outcome.

Cambridge Neuropsychological Test Automated Battery

Neuropsychological testing can be used for the early detection of cognitive deficits, such as episodic memory, across a range of neuropsychiatric disorders, including Alzheimer's disease. For instance, the Cambridge Neuropsychological Test Automated Battery (CANTAB) Visuospatial Paired Associates Learning (PAL) test is a sensitive measure of episodic memory[13] that has been shown to engage the hippocampal formation.[14] The test is very sensitive to mild cases of Alzheimer's disease, and can detect the change from mild cognitive impairment to Alzheimer's disease. The PAL test used in the clinic is comprised of one-pattern, two-pattern, three-pattern, six-pattern, and eight-pattern stages. Studies from our laboratory have shown that, compared with healthy controls, patients with mild cognitive impairment (MCI) show significantly increased hippocampal activation at lower levels of task difficulty and significantly decreased activation at higher levels of task difficulty.[15] The increased activation at lower levels of difficulty shown by MCI patients reflects brain effort in an attempt to perform well on the task. At higher levels of difficulty, the task becomes too challenging, and the decreased activation seen represents

[13] J. H. Barnett, T. W. Robbins, V. C. Leeson, B. J. Sahakian, E. M. Joyce, and A. D. Blackwell, 2010, Assessing cognitive function in clinical trials of schizophrenia, *Neuroscience and Biobehavioral Reviews* 34(8), 1161–1177.

[14] M. de Rover, V. A. Pironti, J. A. McCabe, J. Acosta-Cabronero, F. S. Arana, S. Morein-Zamir, and B. J. Sahakian, 2011, Hippocampal dysfunction in patients with mild cognitive impairment: a functional neuroimaging study of a visuospatial paired associates learning task, *Neuropsychologia* 49(7), 2060–2070.

[15] *Ibid.*

failure to meet the demands of the task. Thus, the visuospatial PAL is a model system for developing symptomatic treatments for episodic memory problems. The aim is to develop a drug which would treat the episodic memory problems of MCI patients and normalise their activation patterns so that they were similar to those of healthy elderly controls.

The CANTAB PAL test is now available on an iPad® and is used in over 260 GP practices throughout the United Kingdom (see CANTAB Mobile; www.cambridgecognition.com). The easy-to-use cognitive assessment tool assists GPs in their decision-making and enables early intervention for mild cognitive impairment and dementia. In addition, data collected countrywide can be aggregated anonymously, allowing better planning of healthcare services. It is hoped that cognitive measures provided by CANTAB PAL, used alongside other biomarkers, such as positron emission tomography (PET), functional neuroimaging (fMRI), cerebrospinal fluid, and blood-based markers, will make it possible to target treatment in the early 'pre-clinical' phase before symptoms of dementia have been experienced.

Cutting across Diagnostic Categories

Focusing on cognitive symptoms for treatment is useful because these often cut across traditional diagnostic categories. For instance, working memory is a cognitive function that is impaired in children with attention deficit hyperactivity disorder (ADHD)[16] and in patients with a first episode of psychosis. Working memory is an important cognitive function that forms a core component of most higher-level executive functions, such as planning and problem solving. It allows us to keep information in mind whilst working on a task. We use this ability daily, for example when verbally rehearsing a new telephone number while we are trying to find a pen and paper to write it down, or when using a passcode to access a computer network. We also use it when mentally organising the most efficient order of activities and re-ordering some of them to get the

---

[16] B. J. Sahakian, 2014, What do experts think we should do to achieve brain health?, *Neuroscience and Biobehavioral Reviews* 43, 240–258.

most out of our work schedule for the day. Working memory has a strong relationship with fluid or creative intelligence and also to crystallised intelligence or intelligence quotient (IQ).[17] Correlation studies support a close relationship between working memory and science achievement, and working memory measured at the start of formal education is a more powerful predictor of subsequent academic success than IQ.[18] Therefore, good working memory is essential to ensure good functional outcome and higher levels of cognitive function are associated with greater levels of wellbeing. Furthermore, a treatment for working memory problems may be useful for improving cognition and functional outcome both in ADHD and in first-episode schizophrenia.[19]

Pharmacological Cognitive Enhancement

Cognitive-enhancing drugs, or 'smart drugs', are needed for enhancing cognitive performance and improving the quality of life in patients with neuropsychiatric disorders. Drugs such as methylphenidate, atomoxetine, and cholinesterase inhibitors are already used to treat cognitive dysfunction in disorders such as ADHD and in Alzheimer's disease.[20] However, the development of novel drugs to treat cognitive dysfunction in other disorders, including depression, schizophrenia, and mild cognitive impairment, remains an unmet need. Recently, the novel antidepressant vortioxetine has been demonstrated in clinical trials to treat cognitive symptoms of psychomotor slowing in depression.[21]

[17] N. P. Friedman, A. Miyake, R. P. Corley, S. E. Young, J. C. Defries, and J. K. Hewitt, 2006, Not all executive functions are related to intelligence, *Psychological Science* 17(2), 172–179.

[18] T. P. Alloway and R. G. Alloway, 2010, Investigating the predictive roles of working memory and IQ in academic attainment, *Journal of Experimental Child Psychology* 106(1), 20–29.

[19] Beddington *et al.*, The mental wealth of nations.

[20] Sahakian *et al.*, The impact of neuroscience on society; L.-S. C. d'Angelo, G. Savulich, and B. J. Sahakian, 2017, Lifestyle use of drugs by healthy people for enhancing cognition, creativity, motivation and pleasure, *British Journal of Pharmacology* 174(19), 3257–3267.

[21] R. S. McIntyre, J. Harrison, H. Loft, W. Jacobson, and C. K. Olsen, 2016, The effects of vortioxetine on cognitive function in patients with major depressive

Modafinil, a wakefulness-promoting agent used in the treatment of narcolepsy, has received considerable attention due to its increasing 'off-label' use by healthy individuals for non-medical 'lifestyle' reasons. However, recent evidence indicates that modafinil holds great potential in treating cognitive problems across a range of psychiatric disorders, including depression and schizophrenia. For instance, in acute studies, modafinil improved spatial working memory in patients with a first episode of psychosis[22] and in patients recovering from depression[23] when given in combination with antipsychotic medication or antidepressants, respectively. Modafinil has also been shown to improve cognitive flexibility in patients with schizophrenia.[24] A meta-analysis also revealed that modafinil, taken in combination with antidepressants, reduced the severity of depression.[25] Importantly, the cognitive improvements following modafinil treatment were associated with improved psychosocial functioning.[26] Thus, modafinil may be a promising new treatment that holds great potential in helping patients to achieve better functional outcomes, such as work and independent living. However, large-scale, longer-term studies are needed to determine this.

disorder: a meta-analysis of three randomized controlled trials, *The International Journal of Neuropsychopharmacology* 19(10), 27312740.

[22] L. Scoriels, J. H. Barnett, P. K. Soma, B. J. Sahakian, and P. B. Jones, 2012, Effects of modafinil on cognitive functions in first episode psychosis, *Psychopharmacology* 220(2), 249–258.

[23] M. K. Kaser, J. B. Deakin, A. Michael, C. Zapata, R. Bansal, D. Ryan, *et al.*, 2017, Modafinil improves episodic memory and working memory cognition in patients with remitted depression: a double-blind, randomized, placebo controlled study, *Biological Psychiatry: Cognitive Neuroscience and Neuroimaging* 2(2), 115–122.

[24] D. C. Turner, L. Clark, E. Pomarol-Clotet, P. McKenna, T. W. Robbins, and B. J. Sahakian, 2004, Modafinil improves cognition and attentional set shifting in patients with chronic schizophrenia, *Neuropsychopharmacology* 29(7), 1363–1373; J. Lees, P. G. Michalopoulou, S. W. Lewis, S. Preston, C. Bamford, T. Collier, *et al.*, 2017, Modafinil and cognitive enhancement in schizophrenia and healthy volunteers: the effects of test batter in a randomised controlled trial, *Psychological Medicine* 47(13), 2358–2368.

[25] A. J. Goss, M. Kaser, S. G. Costafreda, B. J. Sahakian, and C. H. Y. Fu, 2013, Modafinil augmentation therapy in unipolar and bipolar depression: a systematic review and meta-analysis of randomized controlled trials, *The Journal of Clinical Psychiatry* 74(11), 1101–1107.

[26] Kaser *et al.*, Modafinil improves episodic memory and working memory cognition in patients with remitted depression.

Importantly, modafinil improves not only 'cold' (emotion-independent) cognition (e.g. attention and planning), but also 'hot' (emotion-laden) cognition, such as emotional facial recognition. For instance, modafinil improved emotional processing in first-episode psychosis.[27] This is important because impairments in 'hot' cognition, such as disturbances in emotional and motivational processes, are a core feature of many mental health disorders and remain complex barriers to treatment entry and engagement. Therefore, novel treatments targeting impairments in 'hot' cognition may be effective for improving functional outcomes, as well as reducing symptoms or motivational deficits. Neuropsychological test batteries are vital tools for assessing the efficacy of treatment in neuropsychiatric disorders. Recently, the 'EMOTICOM' test battery was developed to offer an objective and comprehensive assessment of 'hot' cognitive functions across a range of mental health disorders.[28] The EMOTICOM test battery, which was recently validated in 200 healthy participants, measures emotion processing, motivation, impulsivity, and social cognition. Together with existing tests such as CANTAB, it is anticipated that EMOTICOM will improve mental health research through facilitating and enhancing treatment development and evaluation across a broad range of neuropsychiatric disorders.

Non-pharmacological Cognitive Enhancement

Despite the need for cognitive-enhancing drugs, there has been a recent withdrawal of some drug companies, including UK-based ones, from the development of new drugs for the treatment of psychiatric disorders.[29]

---

[27] L. Scoriels, J. H. Barnett, G. K. Murray, S. Cherukuru, M. Fielding, F. Cheng, and P. B. Jones, 2011, Effects of modafinil on emotional processing in first episode psychosis, *Biological Psychiatry* 69(5), 457–464; J. P. Roiser and B. J. Sahakian, 2013, Hot and cold cognition in depression, *CNS Spectrums* 18(3), 139–149.

[28] A. R. Bland, J. P. Roiser, M. A. Mehta, T. Schei, H. Boland, D. K. Campbell-Meiklejohn, *et al.*, 2016, EMOTICOM: a neuropsychological test battery to evaluate emotion, motivation, impulsivity, and social cognition, *Frontiers in Behavioral Neuroscience* 10(25), 26941628.

[29] G. Miller, 2010, Is pharma running out of brainy ideas?, *Science (New York, N.Y.)*, 329(5991), 502–504; T. R. Insel, B. J. Sahakian, V. Voon, J. Nye, V. J. Brown, B. M. Altevogt, *et al.*, 2012, Drug research: a plan for mental illness, *Nature* 483, 269.

Fortunately, the explosion of technological advances is leading to innovative, non-pharmacological means to improve cognition, functionality, and wellbeing in patients. For instance, cognitive training is an important technique that stimulates learning and adaptive neuroplastic changes, leading to improved functioning of neural networks.[30] Thus far, cognitive training has been shown to improve cognitive performance both in patients with schizophrenia (improvement of memory deficits)[31] and in individuals with mild cognitive impairment.[32] Importantly, cognitive improvements following training are associated with improved functional outcomes (e.g. work, independent living).[33] Yet, despite the potential benefits of cognitive training, the cost and inconvenience of delivery (e.g. need for specialised equipment, participant travel, supervision), in addition to motivational deficits associated with neuropsychiatric symptoms, lead to considerable dropout.[34]

The recent development of gamified cognitive training using 'serious' games on mobile phones and tablet devices has transformed cognitive training from boring and repetitive into a fun and enjoyable activity. Unlike traditional cognitive training paradigms, computer games can be custom-made to be enjoyable, attention-grabbing, and easily accessible, and may thus comprise an appealing treatment option for enhancing cognition in patients. The difficulty level can be adjusted for the individual,

[30] M. S. Keshavan, S. Vinogradov, J. Rumsey, J. Sherrill, and A. Wagner, 2014, Cognitive training in mental disorders: update and future directions, *The American Journal of Psychiatry* 171(5), 510–522.

[31] B. Kluwe-Schiavon, B. Sanvicente-Vieira, C. H. Kristensen, and R. Grassi-Oliveira, 2013, Executive functions rehabilitation for schizophrenia: a critical systematic review, *Journal of Psychiatric Research* 47(1), 91–104; T. Wykes, V. Huddy, C. Cellard, S. R. McGurk, and P. Czobor, 2011, A meta-analysis of cognitive remediation for schizophrenia: methodology and effect sizes, *The American Journal of Psychiatry* 168(5), 472–485.

[32] H. Li, J. Li, N. Li, B. Li, P. Wang, and T. Zhou, 2011, Cognitive intervention for persons with mild cognitive impairment: a meta-analysis, *Ageing Research Reviews* 10(2), 285–296; G. Savulich, T. Piercy, C. Fox, J. Suckling, J. B. Rowe, J. T. O'Brien, and B. J. Sahakian, 2017, Cognitive training using a novel memory game on an iPad in patients with amnestic mild cognitive impairment, *International Journal of Neuropsychopharmacology* 20(8), 624–633.

[33] R. Cavallaro, S. Anselmetti, S. Poletti, M. Bechi, E., Ermoli, F., Cocchi, and E. Smeraldi, 2009, Computer-aided neurocognitive remediation as an enhancing strategy for schizophrenia rehabilitation, *Psychiatry Research* 169(3), 191–196.

[34] Sahakian, *et al.*, 2015, The impact of neuroscience on society: Savulich *et al.*, Focusing the neuroscience and societal implications of cognitive enhancers.

thus making it challenging while keeping motivation high. Furthermore, there is no stigma associated with this form of cognitive training since almost everyone plays games.

'Use It or Lose It'

Research evidence suggests that interventions such as learning, exercise, or cognitive training activate neural networks in the brain. In rats, both learning and physical activity have been shown to increase neurogenesis in the brain.[35] In healthy humans, 14 hours of cognitive training of working memory over five weeks was associated with increased activation in the working memory neural network, as well as changes in dopamine D_1 receptor density in the brain.[36] A series of elegant imaging studies in taxi drivers in London, who have to remember the location of places in space, showed that the size of the hippocampus was related to the time they had spent as a taxi driver.[37] All of these studies support the concept of 'use it or lose it' – the idea that keeping the brain stimulated (through being mentally active) is essential to ensure good cognitive function and protect against age-related decline.[38] Harnessing this evidence, it is possible to develop novel, innovative cognitive-enhancing technologies, such as cognitive training apps on phones or iPads. Gamified cognitive training represents an innovative way for individuals to maintain and improve good brain health and motivation at an earlier stage and throughout life.

[35] E. Gould, A. Beylin, P. Tanapat, A. Reeves, and T. J. Shors, 1999, Learning enhances adult neurogenesis in the hippocampal formation, *Nature Neuroscience* 2(3), 260–265, https://doi.org/10.1038/6365; A. K. Olson, B. D. Eadie, C. Ernst, and B. R. Christie, 2006, Environmental enrichment and voluntary exercise massively increase neurogenesis in the adult hippocampus via dissociable pathways, *Hippocampus* 16(3), 250–260.
[36] T. Klingberg, 2010, Training and plasticity of working memory, *Trends in Cognitive Sciences* 14(7), 317–324.
[37] E. A. Maguire, D. G. Gadian, I. S. Johnsrude, C. D. Good, J., Ashburner, R. S. J. Frackowiak, and C. D. Frith, 2000, Navigation-related structural change in the hippocampi of taxi drivers, *Proceedings of the National Academy of Sciences of the USA* 97(8), 4398–4403.
[38] M. Orrell and B. Sahakian, 1995, Education and dementia, *BMJ: British Medical Journal* 310(6985), 951–952.

Gamified Cognitive Training for Schizophrenia

Schizophrenia is a chronic mental health condition characterised by positive symptoms, including delusions and hallucinations, as well as cognitive and motivational deficits (negative symptoms). Whereas positive symptoms are reasonably well treated by current antipsychotic medication, the treatment of both cognitive and motivational deficits remains an unmet need. Cognitive deficits are typically found in memory, cognitive flexibility, and visuospatial learning, and are related to functional outcome,[39] meaning that they frequently prevent patients from returning to university or work. Added to this are significant motivational deficits, which also represent a huge barrier to rehabilitation. Schizophrenia is estimated to cost £13.1 billion per year in total in the UK; thus, even small improvements in cognitive functions could help patients make the transition to independent living and working.

We previously demonstrated that gamified cognitive training can improve episodic memory and functional outcome in patients with schizophrenia.[40] In this study, we developed a novel memory game, 'Wizard' (Figure 6.2), for delivering cognitive training of episodic memory using a hand-held portable iPad. The game was based on neuropsychological and neuroimaging evidence. It was piloted and refined on the basis of feedback from a user group of people with schizophrenia. The game was designed to be fun, attention-grabbing, motivating, and easy to understand, whilst at the same time having the potential to improve the player's episodic memory by activating the hippocampal network. It embedded a memory task into a narrative, which allowed selection of characters, rewarded progress, provided feedback, and used visually appealing displays and stimulating music to keep users engaged and motivated. It was found that eight hours of cognitive training (i.e. gameplay) improved episodic memory (patients made fewer errors at the most difficult stage of the CANTAB PAL) and daily functioning in individuals with schizophrenia

---

[39] J. H. Barnett, B. J. Sahakian, U. Werners, K. E. Hill, R. Brazil, O. Gallagher, and P. B. Jones, 2005, Visuospatial learning and executive function are independently impaired in first-episode psychosis, *Psychological Medicine* 35(7), 1031–1041.

[40] Sahakian *et al.*, The impact of neuroscience on society.

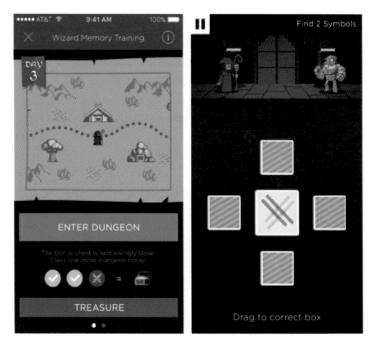

FIGURE 6.2 'Wizard'. Images from 'Wizard', an example of gamified cognitive training of episodic memory, which embedded a learning and memory task into a narrative, allowed selection of characters, rewarded progress, provided feedback, and used visually appealing displays and stimulating music to keep users engaged and motivated (with permission from Peak; www.peak.net).

(increased global assessment of functioning (GAF) score). Importantly, participants in the cognitive training group enjoyed playing the game and were motivated to continue playing across the eight hours of cognitive training (Figure 6.3).

These data suggest that cognitive training using games may serve to complement current pharmacological treatments and/or psychological therapies for schizophrenia. Gamified cognitive training may also have the potential for use in other groups, such as healthy elderly individuals or patient groups with memory-related difficulties (e.g. mild cognitive impairment, traumatic brain injury). The recent adaptation of the Wizard game as an app for mobile phones[41] will ensure access and availability.

---

[41] www.peak.net/advanced-training [accessed 6 August 2018].

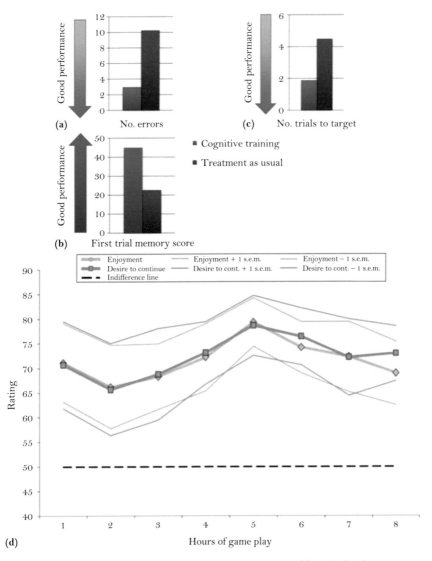

FIGURE 6.3 Effects of 'Wizard' on cognition and functioning in schizophrenia. Memory game training in schizophrenia: the cognitive training group made fewer errors (a), had an improved memory score (b), and had better functioning (c) at outcome. There were no significant differences in these three measures (a, b, c) at baseline. (d) The cognitive training group indicated that they enjoyed playing the game and were motivated to continue across all the hours of cognitive training (all ratings higher than 65%). Modified from Sahakian *et al.* (2015) The impact of neuroscience on society: cognitive enhancement in neuropsychiatric disorders and in healthy people. *Philos. Trans. R. Soc. Lond. B: Biol. Sci.* 370(1677), 20140214, Copyright (2015), with permission from the Royal Society (http://rstb.royalsocietypublishing.org/content/370/1677/20140214). CC BY 4.0 (https://creativecommons.org/licenses/by/4.0/legalcode).

Gamified Cognitive Training for Professional Athletes Susceptible to Head Injury

There has been much concern over concussions and traumatic brain injuries and their effects on cognition and health in the process of playing professional sports, including rugby and NFL football.[42] A systematic review and meta-analysis in Rugby Union concluded that concussion is a common injury sustained during match play, and to a lesser extent in practice, by Rugby Union players.[43] We previously studied the cognitive and emotional effects of brain injury acquired from a variety of sources, and found that survivors of head injury show a diverse pattern of psychiatric and cognitive profiles, including deficits in attention, learning, and memory, as well as slowed processing speed. In a recent joint initiative bringing together the Northampton Saints Rugby Club, Peak, and the Sahakian laboratory, the effects of playing a memory game on cognition in professional rugby players were investigated. Athletes played the University of Cambridge and Peak Memory Advanced Training Plan[44] for up to eight hours, and were tested before and after gameplay sessions. Though only a small sample of rugby players were tested and the athletes played the memory game for different amounts of time, there was a significant relationship between performance on the CANTAB PAL test and the amount of time they played the game (i.e. those who played for longer made fewer errors and needed fewer trials to be successful; Figure 6.4). These data are encouraging, although a much larger study needs to be conducted to replicate and extend these preliminary findings. It is also impressive that the Northampton Saints Rugby Club is concerned and supportive to ensure the brain health and wellbeing of their players. It is notable that in the USA, a much larger but similar approach is being taken by The Football Players Health Study at Harvard University, which is funded by the National Football League Players Association (NFLPA).[45]

[42] 'Shontayne Hape: my battle with concussion', www.nzherald.co.nz/nz/news/article.cfm?c_id=1&objectid=11264856.
[43] A. J. Gardner, G. L. Iverson, W. H. Williams, S. Baker, and P. Stanwell, 2014, A systematic review and meta-analysis of concussion in rugby union, *Sports Medicine (Auckland, N.Z.)* 44(12), 1717–1731.
[44] www.peak.net/advanced-training [accessed 6 August 2018].
[45] https://footballplayershealth.harvard.edu/ [accessed 6 August 2018].

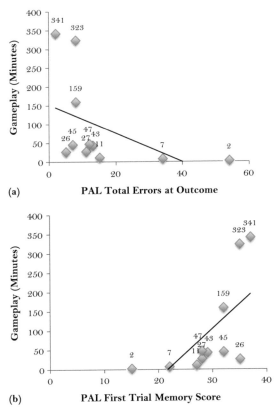

(a)   **PAL Total Errors at Outcome**

(b)   **PAL First Trial Memory Score**

FIGURE 6.4 Effects of cognitive training (gameplay) on cognition and functioning in rugby players. (a) The association between minutes of gameplay by the rugby players and the total error score on the CANTAB Paired Associate Learning test (CANTAB PAL) at outcome (second test). Spearman's $\rho = -0.72$, $p = 0.013$. (b) The association between minutes of gameplay by the rugby players and the first trial memory score on the CANTAB PAL at outcome (second test). Spearman's $\rho = 0.79$, $p = 0.004$.

Conclusion

Harnessing the potential of novel technologies will improve functioning, quality of life, and wellbeing in psychiatric and neurodegenerative disorders, as well as in healthcare in general. Whilst cognitive-enhancing drugs are beneficial for some patients with psychiatric and neurodegenerative disorders, games for cognitive training could be given in combination with pharmacological treatments to achieve the best outcome. For instance, combining novel techniques, such as games, with cognitive-enhancing drugs may promote maximum plasticity for learning. In addition, the combination of treatments also has the potential to improve treatment compliance of patients, for example, through increasing motivation and possibly also by reducing medication dose and associated problematic side effects. The widespread use of gaming technology could also help to reduce some of the stigma associated with mental health treatments.

Healthy individuals should also strive to ensure good brain health and mental wellbeing throughout the lifespan. In the same way as we look after our bodies through healthy eating, exercise, and monitoring and by detecting the onset of physical illness through health check-ups and screening, individuals should also be attentive to their brain health and mental wellbeing. Cognitive function is associated with better wellbeing, and there is huge scope for improving mental capital and wellbeing through different types of intervention.[46] Well-established ways to enhance cognition in healthy humans include education and physical exercise.[47] Both learning and exercise can have a direct effect on mental health and wellbeing across all age groups and should continue throughout life. For instance, exercise improves mathematical and reading achievement in children aged 9–10 years, and also improves cognition and increases life expectancy in healthy older adults.[48] In addition to exercise and lifelong learning, it is important to have a healthy diet and get enough good-quality sleep. Finally, the Foresight Project on Mental Capital and Wellbeing reported on five evidence-based ways to improve personal wellbeing, which include building social networks, physical activity, mindfulness, lifelong learning, and charitable giving.

[46] https://assets.publishing.service.gov.uk/government/uploads/system/uploads/attachment_data/file/292453/mental-capital-wellbeing-summary.pdf [accessed 6 August 2018]; Beddington *et al.* The mental wealth of nations.

[47] K. I. Erickson, C. H., Hillman, and A. F. Kramer, 2015, Physical activity, brain, and cognition, *Current Opinion in Behavioral Sciences* 4, 27–32.

[48] S. Colcombe and A. F. Kramer, 2003, Fitness effects on the cognitive function of older adults: a meta-analytic study, *Psychological Science* 14(2), 125–130; J. F. Sallis, T. L. McKenzie, B. Kolody, M. Lewis, S. Marshall, and P. Rosengard, 1999, Effects of health-related physical education on academic achievement: project SPARK, *Research Quarterly for Exercise and Sport* 70(2), 127–134.

Mental health, like good physical health, has to be worked at. This will require active participation by society and government. However, by using neuroscience, innovation, and technology, we can achieve good brain health and wellbeing across the lifespan for all members of society.

Acknowledgements. This work was supported by the NIHR MedTech and in vitro Diagnostic Co-operative and the NIHR Cambridge Biomedical Research Centre (BRC) (Mental Health Theme and Neurodegeneration Theme).

7 Games Animals Play

NICHOLAS B. DAVIES

Introduction

Games are played whenever the best thing for one individual to do depends on what everyone else is doing. Under this broad definition, we have all been playing games every day of our lives. Our choice of friends and partners depends not only on whether we like them, but also on whether they like us. The way we earn and spend our money depends on what others do. Moment-to-moment decisions depend on the choices of others, too, for example which queue to join in traffic or when shopping, or where to sit in a lecture hall. We often check where everyone else is and what they are doing, before we decide what it is best to do ourselves.

Animals face similar problems in a ruthless natural world, full of competitors and enemies. Their choices of where to go, what to eat, and who to trust also depend on the actions of other individuals, and these choices will influence their success in the game of life. Their goal is simple, and it is selected naturally during evolution, namely to pass on more copies of their genes to future generations than their competitors. For many animals, this is achieved by producing as many healthy offspring as possible. This need not be a conscious decision, of course. It is just inevitable that the world will be populated by individuals who are good at doing this and so their successful traits will be the ones passed on to future generations. This naturally selected goal is so clear that it is often easier to analyse animal games than human games, because with humans it is not always obvious what we are trying to achieve.[1]

---

[1] J. Maynard Smith, *Evolution and the Theory of Games* (Cambridge: Cambridge University Press, 1982).

I shall discuss two kinds of games: *behavioural games*, where individuals make moment-to-moment decisions, adjusting their tactics in response to what others do, and *evolutionary games*, where the strategies change over the generations, as new tricks evolve to outcompete and replace old ones. Long ago, Darwin realised that the outcome of individual competition produced some of the wonders of the natural world. In the final paragraph of *The Origin of Species* he likened nature to 'an entangled bank' in which a struggle for life and, as a consequence, natural selection, led to the evolution of 'endless forms most beautiful and most wonderful'.[2] However, it is only in the last 40 years that we have begun to appreciate how natural selection can lead to remarkable variation within animal populations. The key is to focus on the economics of the decisions that animals make, which involves trade-offs between the costs and benefits of different choices. We then need to consider not only how animals have to adapt to their ecological environment (climate, habitat, enemies, food and other resources necessary for survival and reproduction) but also how they might best cope in a competitive social environment. It is here that intense games are played as individuals are selected to beat their rivals. The outcomes of evolution can be extraordinary; a world not only with extreme cruelty and outrageous cheating but also one with astonishing beauty, too.

Competing for Mates: Charmers, Bullies, and Cheats

I begin with games involving competition for mates. A study of natterjack toads, *Bufo calamita*, by Anthony Arak is a wonderful introduction to a world where the success of different male tactics depends on what others are doing and one which can lead to a mix of charmers and cheats.[3] In spring, males assemble at ponds where females come to lay their eggs, and they call to attract a mate. The result is a chorus of competing males in which females tend to approach those with the loudest calls. A chosen male then climbs on the female's back, and he fertilises her eggs as they

---

[2] C. Darwin, *The Origin of Species* (London: John Murray, 1859).
[3] A. Arak, 1983, Sexual selection by male–male competition in natterjack toad choruses, *Nature* 306, 261–262; A. Arak, 1988, Callers and satellites in the natterjack toad: evolutionarily stable decision rules, *Animal Behaviour* 36, 416–432.

FIGURE 7.1 Charmers and cheats: a calling male natterjack toad with a satellite male waiting silently nearby. Satellites attempt to intercept females attracted to calling males. Photo: Nick Davies.

are laid. However, not all the males are 'callers'. Some are silent 'satellites' who try to creep up and sit next to a caller (Figure 7.1). If they are detected, the caller chases them off, and with good reason, because, when a female approaches, the satellite tries to intercept her.

If the calling males are removed temporarily, and kept in a bucket, then the satellites soon begin to call. So why did they choose to remain silent? Measurements reveal that satellites are small males with weaker voices, who would be unlikely to compete successfully in a loud chorus. Furthermore, they choose to sit next to males who have much louder voices than their own. By intercepting a share of the females of a more attractive male, they do better than they would if they had called themselves.

Overall, calling is the more successful tactic for male natterjack toads, and it is likely that satellite males change tactics and become callers as they get older and larger. In other cases, however, parasitising the efforts of more charming males involves such specialised morphology and behaviour that the difference between males is determined genetically and so a male's strategy is fixed for life. A shorebird, the ruff, *Philomachus*

pugnax, provides an example.[4] The scientific name signifies 'love of fighting', and this is a good description of one of the male strategies. These
'territorial' males have dark neck ruffs, and they are very aggressive,
defending small mating territories on leks, display grounds where males
aggregate to compete for mates. Other males have a white neck ruff, like
a flag of surrender. They do not fight, but rather act as 'satellites' on the
edge of territories, where they sneak copulations, for example while a
territorial male is busy chasing an intruder. A third genetic morph has
been discovered recently, in which some males are small and lack neck
ruffs altogether, and so look just like females. These 'female mimics' are
also likely to behave as sneakers.[5]

It is not yet clear how these three male genetic strategies persist in
the ruff. Most males are territorial (about 85–90%), with satellites rarer
(5–15%), and female mimics rarer still (1%). It is likely that the success of
each male strategy depends on their frequency in the population. Thus,
if all the males were territorial, sneaks would flourish. But at the opposite extreme, in a population of sneaks, territorial males would do better. Thus, frequency-dependent advantage might lead to a stable mix of
genetic strategies.[6]

These frequency-dependent dynamics are clear in side-blotched lizards, *Uta stansburiana*, where there are also three different genetic male
strategies (Figure 7.2(a)).[7]

[4] C. Küpper, M. Stocks, J. E. Risse, N. dos Remedios, L. F. Farrell, S. B. McRae,
et al., 2015, A supergene determines highly divergent male reproductive morphs
in the ruff, *Nature Genetics* 48, 79–83; S. Lamichhaney, G. Fan, F. Widemo,
U. Gunnarsson, D. S. Thalmann, M. P. Hoeppner, *et al.* 2015. Structural genomic
changes underlie alternative reproductive strategies in the ruff (*Philomachus pugnax*), *Nature Genetics* 48(1), 84–88.
[5] J. Jukema and T. Piersma, 2006, Permanent female mimics in a lekking shorebird,
Biology Letters 2, 161–164; D. B. Lank, L. L. Farrell, T. Burke, T. Piersma, and
S. B. McRae, 2013, A dominant allele controls development into female mimic
male and diminutive female ruffs, *Biology Letters* 9(6), 20130653.
[6] F. Widemo, 1998, Alternative reproductive strategies in the ruff *Philomachus
pugnax*: a mixed ESS?, *Animal Behaviour* 56(2), 329–336.
[7] B. Sinervo and C. M. Lively, 1996, The rock–paper–scissors game and the evolution
of alternative male strategies, *Nature* 380, 240–243; S. H. Alonzo and B. Sinervo,
2001, Mate choice games, context-dependent good genes, and genetic cycles in the
side-blotched lizard, *Uta stansburiana*, *Behavioural Ecology and Sociobiology* 49(2/3),
176–186; A. Corl, A. R. Davis, S. R. Kuchta, and B. Sinervo, 2010, Selective loss
of polymorphic mating types is associated with rapid phenotypic evolution during morphic speciation. *Proceedings of the National Academy of Sciences of the USA*
107(9), 4254–4259.

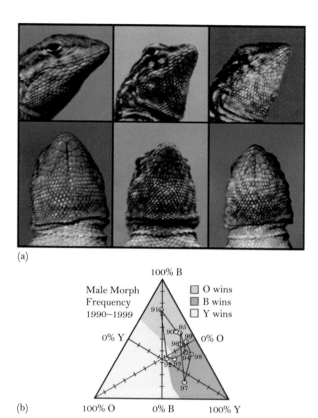

(a)

(b)

FIGURE 7.2 (a) The three male colour morphs of the side-blotched lizard: orange, blue, and yellow. Each has a different mating strategy (see the text). (b) They play an evolutionary game of 'Rock–Paper–Scissors' represented by this triangle. Shaded areas indicate the zones where each morph has the highest fitness. The points plot the observed frequencies of each male strategy (O, B, Y) in a Californian population during the period 1990–1999. The triangle plots the frequencies as follows: 0–100% Blue from base to apex, 0–100% Orange from right side to left vertex, and 0–100% Yellow from left side to right vertex. After Alonso and Sinervo (2001).

> *Orange-throated males* are aggressive and defend large territories with many females.
>
> *Yellow-throated males* look like receptive females (which also have yellow throats). They do not defend territories. Instead, they attempt to sneak matings.
>
> *Blue-throated males* are less aggressive than orange-throated males. They defend small territories in which they guard a single female.

Just as in the human game, 'Rock–Paper–Scissors', the best strategy depends on what others do. In a population of Orange males (Paper), Yellow males (Scissors) do well, because there are many opportunities to sneak matings. However, in a population of Yellow males, Blue males (Rock) do well, because they guard their one female closely. And then we repeat the cycle, because in a population of Blue males, Orange males now do well again, as they gain more mates. So the cycle continues. Observations over a decade, of a population of these lizards along a 250-m sandstone ridge in California, revealed yearly changes in frequencies of the different male strategies as expected from this game, in which each strategy has a strength, which enables it to outcompete males of one morph, but also a weakness, which leaves it vulnerable to exploitation by males of another morph (Figure 7.2(b)).

When humans play the Rock–Paper–Scissors game, the stable solution is to play each strategy at random a third of the time. No competitor can then beat you. Why, then, do we not see this as a behavioural game in the lizards? The most likely answer is that the three male lizard strategies entail such different behaviour that it would be impossible for individuals to make such rapid changes. The equivalent in the human game would not be simple, symbolic changes in hand shapes (flat for paper, two open fingers for scissors, clenched fist for rock), where it is easy to switch, but having to fashion effective scissors from paper material and rock from scissors, which is not feasible. The result in the lizards, therefore, is genetic specialisation and a game that is played over the generations.

The main conclusion from these and other recent studies is that competition for mates often leads to a mix of strategies in a population – charmers, bullies, guarders, and sneaks – the result of a game in which whenever some competitors attempt to gain mates in one way, this opens up opportunities for others to exploit them by alternative strategies.[8]

Sperm Competition: A Mating Lottery

Competition for mates was familiar to Darwin, and led to his theory of sexual selection.[9] A hundred years later, Geoff Parker realised that this is only half the story; sexual competition often continues after the act of

---

[8] For more, see N. B. Davies, J. R. Krebs, and S. A. West, *An Introduction to Behavioural Ecology*, 4th edn (Oxford: Wiley-Blackwell, 2012).

[9] C. Darwin, *The Descent of Man and Selection in Relation to Sex* (London: John Murray, 1871).

Nicholas B. Davies

mating as sperm from rival males compete to fertilise a female's eggs.[10] The result is a game of sperm competition.[11]

This game is illuminated by a comparison between gorillas, *Gorilla gorilla*, and chimpanzees, *Pan troglodytes*. Gorillas live in groups in which a single dominant male has control of a harem of females (Figure 7.3(a)). In effect, he has done all the hard work before copulation, by fighting off sexual competitors. In contrast, chimpanzees live in larger groups where many males (sometimes twenty or more) compete for copulations with the females living in their group (Figure 7.3(b)). A female signals her receptivity with conspicuous sexual swellings which last about 12 days each menstrual cycle. During this period, she copulates on average three or four times an hour and with many different males, often five or so, but sometimes as many as twenty to thirty, depending on the size of the group.[12] Dominant males usually gain most of the matings, particularly during a female's peak fertility, and so sire most offspring.[13] Nevertheless, they face intense sperm competition. Male chimpanzees have evolved to be well equipped to compete in this sperm lottery; they are only a third of the body weight of a gorilla, but they have testes four times as large.

A comparison across primate species, from the tiny 320-g marmoset, *Callithrix*, to the 170-kg gorilla, reveals, as expected, that larger primates have larger testes. But, for a given body weight, those living in multi-male groups have much larger testes than those where a single male controls one or more females (Figure 7.3(c)). Testis size has clearly evolved in relation to the demands of a game of sperm competition.[14]

[10] G. A. Parker, 1970, Sperm competition and its evolutionary consequences in the insects, *Biological Reviews* 45(4), 525–567.
[11] T. R. Birkhead, 2000. *Promiscuity: An Evolutionary History of Sperm Competition and Sexual Conflict* (London: Faber and Faber, 2000).
[12] D. P. Watts, 2007, Effects of male group size, parity, and cycle stage on female copulation rates at Ngogo, Kibale National Park, Uganda, *Primates* 48(3), 222–231.
[13] C. Boesch, G. Kohou, H. Néné, and L. Vigilant, 2006, Male competition and paternity in wild chimpanzees of the Taï forest, *American Journal of Physical Anthropology* 130(1), 103–115; N. E. Newton-Fisher, M. E. Thompson, V. Reynolds, C. Boesch, and L. Vigilant, 2010, Paternity and social rank in wild chimpanzees (*Pan troglodytes*) from the Budongo Forest, Uganda. *American Journal of Physical Anthropology* 142(3), 417–428; M. N. Muller, M. E. Thompson, S. M. Kahlenberg, and R. W. Wrangham, 2011, Sexual coercion by male chimpanzees shows that female choice may be more apparent than real, *Behavioural Ecology and Sociobiology* 65(5), 921–933.
[14] A. H. Harcourt, P. H. Harvey, S. G. Larson, and R. V. Short, 1981, Testis weight, body weight and breeding systems in primates, *Nature* 293, 55–57; A. H. Harcourt,

126

(a)

(b)

(c)

FIGURE 7.3 Sperm competition games: a dominant male mountain gorilla (a) defends a harem where he can monopolise matings with his females (photo copyright Stuart Butchart), whereas male chimpanzees (b) live in multi-male groups where a female often mates with many males (photo copyright Kathelijne Koops). (c) Primate testis size reflects differences in the intensity of sperm competition. The graph plots combined testes weight versus body weight for different primate genera. Those with multi-male breeding systems (solid circles, chimps are the uppermost point), and hence intense sperm competition, have larger testes than those where one male can monopolise matings, either in monogamy (open circles) or by defending a harem (open triangles, gorillas are the far right point). The cross marks our own species. After Harcourt *et al.* (1981).

Humans, too, bear the indelible stamp of their evolutionary past. Human testis size is more or less as expected for an average primate of our body weight (Figure 7.3(c)). This suggests that our ancestors were not quite as relaxed about paternity as male gorillas, but not so worried as a male chimpanzee.

Sexual Conflict: Games between the Sexes

Mating and sperm competition games become all the more intriguing when there is sexual conflict, namely different best outcomes for a male and for a female.[15] These conflicts are familiar in our own lives, and they have inspired literature, drama, and art, so it is odd that biologists were slow to recognise this as a powerful selective force in animal lives. One explanation is that we used to avoid anthropomorphism in scientific studies of animal behaviour. But imagining what we might do if we were a chimp or a lizard often provides valuable insights; when we simulate the various costs and benefits of alternatives in our brains, to judge what might be best, we mimic what natural selection is doing over the generations. The first theoretical treatments of sexual conflict in animal societies were by Geoff Parker in the 1970s, and empirical studies have flourished ever since, revealing that males and females often have different best options in all aspects of family life, from mate choice through to parental care.[16]

A game of sexual conflict, played in British gardens every summer, is that of competition for mates in the dunnock (or hedge sparrow), *Prunella modularis* (Figure 7.4(a)). 'Dun' means dull brown and 'ock' signifies little, and the dunnock is the archetype of a little brown bird. In his book *The History of British Birds*, published in 1856, the Reverend F. O. Morris was so impressed with the dunnock's modest appearance that he urged his parishioners to emulate its lifestyle.[17] This is what

A. Purvis, and L. Liles, 1995, Sperm competition: mating system, not breeding season, affects testes size of primates, *Functional Ecology* 9(3), 468–476.
[15] G. A. Parker, 2006, Sexual conflict over mating and fertilization: an overview, *Philosophical Transactions of the Royal Society B: Biological Sciences* 361(1466), 235–259.
[16] G. Arnqvist and L. Rowe, *Sexual Conflict* (Princeton, NJ: Princeton University Press, 2005).
[17] F. O. Morris, *A History of British Birds* (London: Groombridge, 1856).

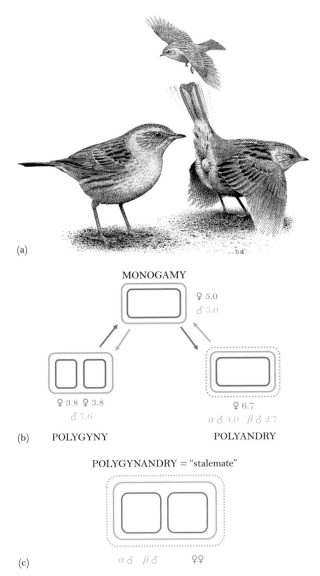

(a)

MONOGAMY

♀ 5.0
♂ 5.0

♀ 3.8 ♀ 3.8
♂ 7.6

♀ 6.7
α♂ 4.0 β♂ 2.7

(b)   POLYGYNY   POLYANDRY

POLYGYNANDRY = "stalemate"

(c)   α♂ β♂   ♀♀

FIGURE 7.4 Dunnock mating games. (a) A female dunnock soliciting to a beta male, while the alpha male flies in to interrupt them. Drawing copyright David Quinn, reproduced with his permission, from Davies (1992). (b) Sexual conflict gives rise to various mating systems; male territories in blue, female territories in red, with seasonal reproductive success (number of chicks raised) indicated for males and females in each system. A female does best with two males (polyandry) while a male does best with two females (polygyny). Arrows show the opposing behavioural preferences of males and females. (c) Polygynandry (two males sharing two females) as a stalemate outcome of sexual conflict. Artwork copyright the author, reproduced from Davies (1992); see the text.

he wrote: 'Unobtrusive, quiet and retiring, without being shy, humble and homely in its deportment and habits, sober and unpretending in its dress, while still neat and graceful, the dunnock exhibits a pattern which many of a higher grade might imitate, with advantage to themselves and benefit to others through an improved example.' Our studies in the Cambridge University Botanic Garden reveal that his recommendation is unfortunate. Had his congregation followed suit, the parish would have been in chaos.

The key to understanding the game is to follow individuals, by marking them with colour bands, so we can watch their behaviour closely, and then to measure their reproductive success by scoring the number of offspring they produce, using DNA markers to assign maternity and paternity.[18] Once individuals are recognisable, a whole new world is revealed, with all the drama and excitement of an avian soap opera. We found various mating arrangements, depending on how male territories overlapped female territories (Figure 7.4(b)). Sometimes one male defended a territory that overlapped one female territory (monogamy; presumably what the Reverend Morris had in mind for dunnocks), or, in other cases, a male territory overlapped two adjacent female territories (polygyny). Sometimes two males shared a territory, either with one female (polyandry) or with two adjacent females (polygynandry, Figure 7.4(c)). The two males were never close relatives, and one (the 'alpha') was always dominant to the other ('beta').

As soon as a female has completed her nest, she begins to solicit copulations. This usually starts a week before she lays her first egg and continues until the clutch is complete, a period of about 10 days (four or five eggs are laid, one per day, in the early morning). Monogamous males follow their female throughout this time and chase off any intruding males. But the most intense conflicts occur when two males share a territory. Here, the alpha male attempts to guard the female, to prevent the beta male from copulating with her. Wherever she goes he follows, usually feeding or perched within a metre or two of her, and ready to chase off the beta male whenever he approaches. However, he faces a difficult task,

---

[18] N. B. Davies, *Dunnock Behaviour and Social Evolution* (Oxford: Oxford University Press, 1992).

not only because beta males are often very persistent, but also because the female frequently goes to the beta male in order to solicit to him (Figure 7.4(a)). This provokes endless chases around the territory. When the female flies off, both males follow in hot pursuit. Sometimes, the female escapes the alpha male's guarding by flying off suddenly, and then she hides away with the beta male and can copulate with him in peace. Meanwhile the alpha male searches frantically round the territory, and as soon as he finds them the chases start all over again. A female with two males may copulate several times an hour, as they battle for paternity.

Just like chimpanzees, dunnocks have unusually large testes to fuel such frequent copulation. But there is also an extraordinary display prior to mating, which is likely to enhance a male's success in the sperm lottery. The male stands behind the female and pecks her cloaca. After a series of cloacal contractions, there is a brief spasm when the female suddenly dips her abdomen and ejects a droplet. Then the male copulates very quickly, with a brief cloacal 'kiss' as he jumps over her. These ejected droplets contain sperm from previous matings, probably from the cloaca itself, and so it is likely that the display helps to ensure a clear passage for the male's insemination.

Like other birds, female dunnocks have sperm stores, and they mobilise sperm from their stores to fertilise each egg in turn, about 24 hours before it is laid. Our DNA markers revealed that shared mating with two males often led to mixed paternity in the brood. On average, alpha males gained 60% of the paternity and beta males 40%. Males apparently did not recognise their own chicks. Sometimes, by chance, one male (more often the beta) lost out in the sperm lottery and the other male sired the whole brood. Nevertheless, the failed male still helped to feed the chicks provided he got a share of the matings. Furthermore, when there was shared paternity there was no tendency for each male to prefer to feed his own chicks. Males therefore used mating success as an indirect paternity cue. If they failed to mate, they did not help, but if they succeeded they did so.

When we calculated male and female reproductive success in the different mating systems, this made beautiful sense of the behavioural conflicts we observed (Figure 7.4(b)). The success of a brood depended on how well it was fed, with the more providers the better. This meant that a female had least success in polygyny, where she had only the part-time help of a male, because he spent half his time, on average, helping his other female to raise her brood. She did better with monogamy, where she had a male's full-time help, and best of all in polyandry, where she had full-time help from two males, provided she had copulated with them both. These pay-offs explain why a polygynous female tried to drive the other female away, to claim the male's full-time help for her own brood. And they also explain why a female was so keen to give a mating share to a beta male, to secure help from two males rather than one.

A male, however, did best with the opposite mating arrangement. Although more chicks were raised in polyandry, the system where a female did best, here a male did worst because of the cost of shared paternity. A male did better in monogamy, where he enjoyed full paternity of a smaller brood. And he did best of all in polygyny, where a female did worst, because he gained full paternity of the sum of two broods.

These conflicting pay-offs, resulting from the cost of polygyny to a female from shared male care and the cost of polyandry to a male from shared paternity, explain the sexual conflicts in behaviour. Thus, in polyandry, while it pays a female to encourage a beta male, it pays the alpha male to drive him off. And in polygyny, while the two females squabble, the male hops in between them to try to keep the peace. Polygynandry, with two males sharing two females, reflects a 'stalemate' in the game of sexual conflict; the alpha male is unable to chase the beta male off and claim both females (his best option – polygyny), while neither female can evict the other to claim both males (her best option – polyandry).

Cuckoos versus Hosts: Hide and Seek, Signatures and Forgeries

If birds did not already have sufficient problems with their own mating games, they also face threats from some of nature's most notorious cheats. About 1% of the 10,000 bird species in the world are obligate brood parasites, namely they never raise their own young but instead trick other species into doing all the work. Here, I focus on the common cuckoo, *Cuculus canorus* (henceforth 'the cuckoo'), one of the best studied tricksters, which exploits many species of small birds as its hosts right

FIGURE 7.5 A cuckoo egg (middle, left) in a reed warbler nest. This cuckoo race has evolved an egg that matches the host's eggs, the outcome of an evolutionary game involving host defences versus cuckoo trickery (photo copyright Dave Leech).

across the Palaearctic, from Western Europe to Japan. Recent studies have revealed that cuckoo–host interactions involve both behavioural games of hide and seek and evolutionary games, with escalation of host defences and cuckoo trickery over the generations.[19]

The female cuckoo lays just one egg in each host nest (Figure 7.5). Soon after hatching, the cuckoo chick ejects the host's eggs and chicks; it balances each of them on its back, one by one, climbs up to the rim of the nest, and tosses them overboard with a flick of its wing stumps. Having cleared the nest of all competition, the cuckoo chick then commands all the food the host parents bring to the nest, and so the hosts end up raising a cuckoo chick instead of a brood of their own. In theory, this interaction should provoke an 'evolutionary arms race' in which hosts evolve defences against cuckoos, and cuckoos then respond by evolving better trickery, leading to further escalation of defence and trickery by both parties.[20]

[19] N. B. Davies, *Cuckoo – Cheating by Nature* (London: Bloomsbury, 2015).
[20] R. Dawkins and J. R. Krebs, 1979, Arms races between and within species, *Proceedings of the Royal Society B: Biological Sciences* 205(1161), 489–511.

Experiments with model eggs have indeed revealed a 'ding–dong' battle, in which each side has evolved in response to the other. First of all, hosts tend to reject eggs that differ in appearance from their own eggs. This explains why the cuckoo has evolved into genetically distinct races, each of which specialises on one particular host species and lays an egg type that matches the eggs of its chosen host.[21] In Britain, for example, one race of cuckoo specialises on reed warblers, *Acrocephalus scirpaceus*, and it lays a greenish, spotted egg just like reed warbler eggs, while another specialises on meadow pipits, *Anthus pratensis*, and it lays a brown, spotted egg just like meadow pipit eggs. Experiments show that both these hosts reject eggs unlike their own.

In contrast, the cuckoo race that targets dunnocks lays a pale, spotted egg, nothing like the plain blue eggs of the dunnock. How does this cuckoo race get away with such a bad match? Experiments again provide the answer; dunnocks will accept any coloured egg among their clutch, so there has been no selection for this cuckoo race to evolve a matching egg. Another cuckoo race in Europe that specialises on redstarts, *Phoenicurus phoenicurus*, lays a plain blue egg, a perfect match for this fussy host species. So cuckoos can certainly evolve blue eggs if called to do so. In general, comparing cuckoo races across Europe, the more discriminating the host species, the better the cuckoo's egg mimicry.[22] Therefore, cuckoo trickery clearly evolves in response to the intensity of host defences.

Do host defences, in turn, evolve in response to cuckoo trickery? If so, species of small birds which are unsuitable as hosts should have no defences against cuckoos, because they have no history of cuckoo parasitism. Experiments support this prediction. Species that nest in small holes, inaccessible to female cuckoos, and those that feed their young on a seed diet, unsuitable for raising young cuckoos, accept eggs unlike their own. Therefore, it is likely that the egg-rejection behaviour we observe in

[21] M. de L. Brooke and N. B. Davies, 1988, Egg mimicry by cuckoos *Cuculus canorus* in relation to discrimination by hosts, *Nature* 335, 630–632; H. Lisle Gibbs, M. D. Sorenson, K. Marchetti, M. de L. Brooke, N. B. Davies, and H. Nakamura, 2000, Genetic evidence for female host-specific races of the common cuckoo, *Nature* 407, 183–186; F. Fossøy, M. D. Sorenson, W. Liang, T. Ekrem, A. Moksnes, A. P. Møller, *et al.*, 2016, Ancient origin and maternal inheritance of blue cuckoo eggs, *Nature Communications* 7, 10272.

[22] M. C. Stoddard and M. Stevens, 2011, Avian vision and the evolution of egg colour mimicry in the common cuckoo, *Evolution* 65(7), 2004–2013.

cuckoo hosts evolved specifically in response to cuckoo parasitism. This suggests that the starting point in the arms race is 'no host defence – no cuckoo egg mimicry', and each side then evolves in response to the other. Perhaps dunnocks are recent victims, still at the start of this interaction?

For most cuckoo races, the cuckoo egg is an impressive match of the host's eggs. Hosts are therefore uncertain simply from looking at their clutch whether there is a cuckoo egg present. The result is a game of 'hide and seek'; hosts look out for cuckoos, so they can vary their defences in response to local parasitism risk,[23] and cuckoos counteract this by secrecy, speed, and disguise.

Experiments show that hosts are more likely to reject an egg, or to desert their clutch, if they see a cuckoo at their nest, either a real one or a stuffed cuckoo placed by experiment.[24] However, the female cuckoo is very secretive and her laying visit is sometimes incredibly fast; she can glide down to the host nest, remove a host egg and lay one of her own in its place, all within 10 seconds. It is no surprise, then, that hosts widen their source of information about local cuckoo activity by eavesdropping on the alarms of neighbours; if they are alerted by neighbours to the presence of a cuckoo, they are both more likely to defend their own nests against female cuckoos[25] and to reject cuckoo eggs.[26] Female cuckoos have responded to this by evolving a plumage polymorphism; some females are grey and others are rufous. Experiments with model cuckoos

---

[23] B. G. Stokke, I. Hafstad, G. Rudolfsen, A. Moskmes, A. P. Møller, E. Røskaft, and M. Soler, 2008, Predictors of resistance to brood parasitism within and among reed warbler populations, *Behavioural Ecology* 19(3), 612–620; J. A. Welbergen and N. B. Davies, 2009, Strategic variation in mobbing as a front line of defense against brood parasitism, *Current Biology* 19(3), 235–240; R. Thorogood and N. B. Davies, 2013, Reed warbler hosts fine-tune their defenses to track three decades of cuckoo decline, *Evolution* 67(12), 3545–3555.

[24] N. B. Davies and M. de L. Brooke, 1988, Cuckoos versus reed warblers: adaptations and counteradaptations, *Animal Behaviour* 36(1), 262–284; A. Moskmes, E. Røskaft, L. Greger Hagen, M. Honza, C. Mørk, and P. H. Olsen, 2000, Common cuckoo *Cuculus canorus* and host behaviour at reed warbler *Acrocephalus scirpaceus* nests, *Ibis* 142(2), 247–258.

[25] N. B. Davies and J. A. Welbergen, 2009, Social transmission of a host defence against cuckoo parasitism, *Science* 324(5932), 1318–1320; D. Campobello and S. G. Sealy, 2011, Use of social over personal information enhances nest defense against avian brood parasitism, *Behavioural Ecology* 22(2), 422–428.

[26] R. Thorogood and N. B. Davies, 2016, Combining personal with social information facilitates host defences and explains why cuckoos should be secretive, *Scientific Reports* 6, 19872.

show that, when hosts are alerted to grey cuckoos, rufous cuckoos are more likely to evade host defences, and vice versa.[27] Therefore, by coming in different guises female cuckoos make it harder for hosts to assess the local parasitism risk.

These games of secrecy and detection are not the only ones played by cuckoos and hosts. Their egg patterns evolve, too, in a game of signatures and forgeries. The beautiful patterning on bird egg shells has long been a source of wonder for naturalists and poets alike. At about the time Charles Darwin was finding finches in the Galapagos Islands, John Clare was finding poems in the English countryside. One of his loveliest poems was inspired by the nest of the yellowhammer, *Emberiza citrinella*, also known as the 'scribble lark' on account of the remarkable markings on its eggshells:

> Five eggs, pen-scribbled o'er with ink their shells,
> Resembling writing scrawls, which fancy reads
> As nature's poesy and painted spells.

About 80 years later, a naturalist working in Africa, Charles Swynnerton,[28] suggested that eggshell markings were not just poetic whimsy on the part of nature, but really had evolved as signatures to enable hosts to better recognise their own eggs in the battle with cuckoos. Before Swynnerton, it had been assumed that the colour and spotting of bird's eggs was simply a means of camouflage. Here was a completely new idea, that spots and squiggles were the way the hosts write on their eggs 'this is my egg'. The cuckoo then has to forge the host's signature by writing on its egg 'and so is this'. An evolutionary arms race might then ensue, as hosts evolve new signatures to foil the cuckoo, and cuckoos evolve to keep track with new forgeries.

We now have good evidence that host egg patterns do indeed evolve in response to cuckoos. Compared with species with no history of brood parasitism, cuckoo hosts have more individually distinctive markings on

---

[27] R. Thorogood and N. B. Davies, 2012, Cuckoos combat socially transmitted defenses of reed warbler hosts with a plumage polymorphism, *Science* 337(6094), 578–580.

[28] C. F. M. Swynnerton, 1918, Rejection by birds of eggs unlike their own: with remarks on some of the cuckoo problems, *Ibis* 60(1), 127–154.

their eggs.[29] These 'better signatures' in host species result from two characteristics. First, a given female's egg markings are more replicable. This means that she signs her eggs in the same way every time, so there is less variation between her own eggs. This obviously would make it easier for her to detect a foreign egg in the clutch. Second, the various egg characteristics of different females – for example background colour, size and shape of the spots and squiggles, their spacing and positioning on the egg – all vary independently of one another.[30] This is exactly what we would expect if evolution has maximised the diversity of signatures that are possible. The result is that in host species there is more variation between the eggs of different females. This makes life harder for a cuckoo, because her egg pattern cannot be a good forgery in every nest she parasitises.

Have hosts evolved better signatures as their cuckoos evolve better forgeries? We now have good evidence for this, too. Cassie Stoddard, Rebecca Kilner, and Chris Town[31] devised a computer program which measured the various features of markings on host eggs (shape, size, orientation, and contrast) and then had the task of assigning an egg to its correct clutch. It did this by comparing the number of matches between that egg and other eggs of the same species, and picking the best match. Each host species was then given a 'recognisability score', which was a measure of how well eggs could be assigned to their rightful clutch, in other words a score of how distinct the signatures of individual females were. When different host species were compared, those with the best signatures were those whose cuckoo race had the best match to the host's egg. Therefore, as cuckoos evolve a better match, hosts evolve more individually distinctive signatures.

Nevertheless, the egg signatures of hosts of the common cuckoo are modest compared with those of some cuckoo hosts in Africa, where

[29] B. G. Stokke, A. Moksnes, and A. Røskaft, 2002, Obligate brood parasites as selective agents for the evolution of egg appearance in passerine birds, *Evolution* 56(1), 199–205.

[30] E. M. Caves, M. Stevens, E. S. Iverson, and C. N. Spottiswoode, 2015, Hosts of avian brood parasites have evolved egg signatures with elevated information content, *Proceedings of the Royal Society B: Biological Sciences* 282(1810), 20150598.

[31] M. C. Stoddard, R. M. Kilner, and C. Town, 2014, Pattern recognition algorithm reveals how birds evolve individual egg pattern signatures, *Nature Communications* 5, 4117.

the arms race is likely to be much older. Perhaps it is no surprise that Swynnerton's insight came from the eggs of African birds. The prize for the most spectacular signatures of all must surely go to an African warbler, the tawny-flanked prinia, *Prinia subflava*. This little brown bird is a favourite host of the cuckoo finch, *Anomalospiza imberbis*, whose parasitic habits are like those of the cuckoo. In a wonderful field study in Zambia, Claire Spottiswoode and colleagues have shown that individual females always lay the same egg type, but the variation among different females is astonishing (Figure 7.6). Ground colour varies continuously, from white to red to olive to blue. The markings vary too, from fine spots to large blotches to intricate scribbles. Some females have mainly one type of markings on their eggs, while others have a variety of markings. And the dispersion of markings varies, too, from evenly over the shell to mainly concentrated at the blunt end.

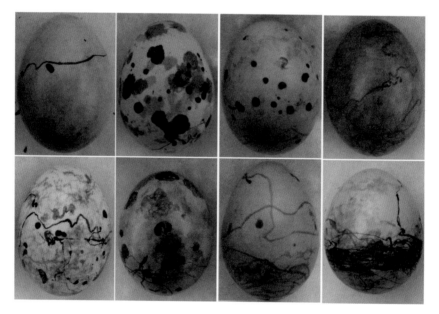

FIGURE 7.6 The evolutionary battle between hosts and their 'cuckoos' can lead to a signature–forgery arms race in egg markings. These are all eggs from one small population of the tawny-flanked prinia (an African warbler) in Zambia. Each egg is from a different female. Individual differences in colour and markings act as signatures by which each female can recognise her own eggs (photo copyright Claire Spottiswoode).

Do the prinias use their individual signatures to help them detect a foreign egg? To test this, Spottiswoode replaced one egg of a clutch with an egg of another female. Most of these were promptly rejected; the nest owner punctured the egg and carried it away. By comparing the characteristics of rejected and accepted eggs, it was found that the prinias paid attention to four signature cues: background colour, marking type, marking variation, and dispersion of the marks. A large difference in just one cue was sufficient to cause rejection, but the prinias integrated information from all four cues, so smaller differences across several cues would lead to rejection too.[32]

Clearly the cuckoo finch would need an outstanding match to fool such particular hosts. So it was thrilling to discover that this parasite had astonishing variation in its eggs too, a set of forgeries which matched, almost perfectly, the range in signatures of its host. Just like the prinias, each female cuckoo finch always lays exactly the same egg type, so ideally she should target only those prinias in the host population for whom her forgery would be a good match. However, the cuckoo finches are not so clever. They parasitised nests haphazardly with respect to host egg type, and as a result many of their eggs were rejected. Only when there was an occasional match, by chance, did the cuckoo finches succeed in tricking the hosts. So the variable signatures of the prinias were a wonderfully effective defence against parasitism.

Is the signature–forgery arms race continuing today? Examination of eggs collected over the past four decades shows that both colours and patterns have changed even over this short time period. Cuckoo finch eggs and host eggs have changed in concert, so parasite eggs are a better match of contemporaneous host eggs. This suggests that parasite forgeries have tracked host signatures through time. However, the tracking is not perfect. Currently there is a rare host egg type, with an olive-green background, that the cuckoo finch has yet to forge. Prinias with this signature have become more common over the past 40 years, presumably

32 C. N. Spottiswoode and M. Stevens, 2010, Visual modelling shows that avian host parents use multiple visual cues in rejecting parasitic eggs, *Proceedings of the National Academy of Sciences of the USA* 107(19), 8672–8676; C. N. Spottiswoode and M. Stevens, 2011, How to evade a coevolving brood parasite: egg discrimination versus egg variability as host defences, *Proceedings of the Royal Society B: Biological Sciences* 278, 3566–3573.

because they are better able to detect any parasite eggs, and so raise more of their own offspring, who inherit the effective signature.[33] This egg type should continue to proliferate until a new forgery appears.

What would happen to egg signatures if a host became freed from parasitism? The ideal test would be to take some individuals of cuckoo host species and release them on an island with no cuckoos. If we followed their descendants over the generations, we would predict that the distinctive egg signatures would gradually diminish because they would no longer be needed. Amazingly, this experiment has been done, but it was done unwittingly and in shameful circumstances. In Africa, village weaverbirds, *Ploceus cucullatus*, are favourite hosts of the diederik cuckoo, *Chrysococcyx caprius*. This cuckoo has a hard time because these hosts are very good at rejecting eggs that are not a perfect match of their own. Just like the prinias, their ability to spot a foreign egg is enhanced by remarkable variation in their egg signatures. An individual female weaverbird always lays one constant egg type, but the pattern varies among females.

During the late eighteenth century, at the peak of the Atlantic slave trade, ships ferrying cargoes of slaves from West Africa to the West Indies sometimes took captive birds on board, too, as pets. Among these were village weaverbirds. Inevitably some of them escaped captivity and today there is a thriving population in Hispaniola. But on this new island home there has been an important change to their lives, because there are no parasitic cuckoos in the West Indies. It is a cruel irony that the journey across the Atlantic led to freedom for these birds but to slavery for their human companions.

David Lahti has compared the eggs of weaverbirds on Hispaniola, free from cuckoos for some 230 years, with those from the ancestral parasitised population in West Africa.[34] As predicted, the clutches in Hispaniola are much less variable both in ground colour and in spotting. Furthermore, individuals now have much more variation within their own clutches. So not only have their signatures become less individually distinct, but also

[33] C. N. Spottiswoode and M. Stevens, 2012, Host–parasite arms races and rapid changes in bird egg appearance, *American Naturalist* 179(5), 633–648.

[34] D. C. Lahti, 2005, Evolution of bird eggs in the absence of cuckoo parasitism, *Proceedings of the National Academy of Science of the USA* 102(50), 18057–18062; D. C. Lahti, 2006, Persistence of egg recognition in the absence of cuckoo brood parasitism: pattern and mechanism, *Evolution* 60(1), 157–168.

they are reproduced less faithfully on their own eggs. The latter finding is particularly interesting, because it could be argued that the introduced populations have less variable eggs simply because they were founded from a small number of individuals who, by chance, did not have the full range of variation in the source population (a sampling effect, known as the 'founder effect'). However, this could not explain why variation within a clutch has increased, which surely reflects an evolutionary loss of signature consistency. Another weaverbird population, introduced in 1886 to the cuckoo-free island of Mauritius in the Indian Ocean, also has less distinctive egg signatures than source populations in Africa. But the changes have been less marked, exactly as we would expect from a more recent introduction, with less time for evolutionary change.

Therefore, we have good evidence not only that hosts evolve better egg signatures in response to cuckoo parasitism, but also that they lose them when parasitism ceases.

A Game of Safety in Numbers: From Local Rules to Group Behaviour

So far, we have seen how the games animals play can lead to variable mating strategies, and strange mixes of cruelty and beauty as cuckoos and hosts do battle over the generations, with new strategies evolving or being lost as selection pressures change. Now we come to an example where, just as in human games, simple rules can produce extraordinary outcomes.

Animal groups provide some of the most remarkable spectacles in the natural world. In winter, as dusk falls, flocks of starlings, *Sturnus vulgaris*, fly from the fields where they have been foraging all day towards their night-time roosts. From all directions they come, congregating into a vast flock, often with tens or hundreds of thousands of individuals wheeling in ever-changing smoke-like formations before they plunge into the safety of the roost (Figure 7.7). In the seas, vast shoals of fish perform similar, spectacular coordinated movements, circling in a tight sphere or suddenly scattering in all directions with flashes of silver.

These spectacles have been a source of wonder to human observers ever since ancient times. Early philosophers imagined that there must be

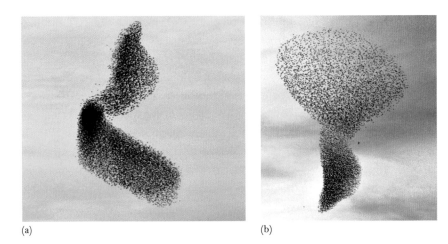

(a)   (b)

FIGURE 7.7 A game of safety in numbers: the spectacular, coordinated movements of tens of thousands of starlings at winter roosts is the outcome of local decisions by individuals in the flock (photos copyright Andrew Smiley).

a leader to orchestrate the game, or perhaps such rapid changes in shape and direction involved 'thought transference' among individuals or other mystical powers. However, recent studies have shown that simple local movement rules, adopted by individuals in response to their neighbours, can produce these amazing outcomes of group behaviour.[35]

If a bird of prey approaches, such as a peregrine, *Falco peregrinus*, starlings respond by joining others and bunching up to form a tight spherical flock. The advantage to an individual of doing this is two-fold.[36] First, flocking dilutes its risk of attack. If it is alone, clearly it is at risk if a predator comes along. If it is in a group of N individuals, it now has a $1/N$ chance of being the victim. As long as there are fewer than N attacks against a group of N potential victims, individuals will be safer in a group through this dilution effect. Therefore individuals may selfishly join others simply in the hope that someone else gets attacked rather than them. Second, an individual will be even safer if it is in the middle of the group, hiding behind a barrier of companions. The result is that the flock is in

[35] I. D. Couzin and J. Krause, 2003, Self-organization and collective behavior in vertebrates, *Advances in the Study of Behavior* 32, 1–75.
[36] W. D. Hamilton, 1971, Geometry for the selfish herd, *Journal of Theoretical Biology* 31(2), 295–311.

constant motion, as individuals jostle for the safest places. When there is a bird of prey at large, the starlings form larger and more compact flocks, and as it attacks waves of escape spread through the flock, as if a giant amoeba were flowing gracefully across the sky.[37]

How can a large flock behave so synchronously, just as if it were one organism? Pioneering studies by William Foster and John Treherne showed how the escape response of one individual could lead to rapid changes in a group.[38] They studied water skaters, *Halobates robustus*, little insects that scuttle over the water surface in search of food and at risk of attack by fish from below and by birds from above. Using a model predator to simulate attacks, they showed that, as soon as one insect made escape movements and bumped into its neighbours, they responded too, and a wave of alarm spread rapidly through the group, so rapidly in fact that those on the far side from the attack were already darting away before they could detect the predator themselves. They named this wave of information the 'Trafalgar effect', after the famous battle where signals were transmitted along a chain of flag ships, allowing Admiral Nelson to know of the movements of the joint French and Spanish fleets, and plan his surprise attack, even when they were still over the horizon and he could not yet detect them himself.

Recent laboratory studies of fish shoals have used computer tracking to follow the movements of individuals in great detail, and have shown how waves of movements can spread rapidly through the group simply as a result of local responses to neighbours.[39] There can be sudden cascades

[37] C. Carere, S. Montanino, F. Moreschini, F. Zoratto, F. Chiarotti, D. Santucci, and E. Alleva, 2009, Aerial flocking patterns of wintering starlings, *Sturnus vulgaris*, under different predation risk, *Animal Behaviour* 77(1), 101–107; A. Procaccini, A. Orlandi, A. Cavagna, I. Giordina, F. Zoratto, D. Santucci, *et al.*, 2011, Propagating waves in starlings, *Sturnus vulgaris*, flocks under predation, *Animal Behaviour* 82(4), 759–765.

[38] W. A. Foster and J. E. Treherne, 1981, Evidence for the dilution effect in the selfish herd from fish predation on a marine insect, *Nature* 295, 466–467; J. E. Treherne and W. A. Foster, 1981, Group transmission of predator avoidance behaviour in a marine insect: the Trafalgar effect, *Animal Behaviour* 29(3), 911–917.

[39] A. Strandburg-Peshkin, C. R. Twomey, N. W. Bode, A. B. Kao, Y. Katz, C. C. Ioannou, *et al.*, 2013, Visual sensory networks and effective information transfer in animal groups, *Current Biology* 23(17), R709–R711; S. B. Rosenthal, C. R. Twomey, A. T. Hartnett, H. S. Wu, and I. D. Couzin, 2015. Revealing the hidden networks of interaction in mobile animal groups allows prediction of complex behavioral contagion, *Proceedings of the National Academy of Sciences of the USA* 112(15), 4690–4695.

of behavioural change in a group as just a few informed individuals flee from an approaching predator[40] or follow others to food.[41] Furthermore, the way information spreads through social networks leads to democratic decisions based on the consensus of uninformed individuals, so the group can often avoid waves of panic that could stem from false alarms or from strongly opinionated or extremist minorities.[42] Perhaps we could learn ourselves from the games animals play.

Acknowledgements. I thank David Blagden and Mark de Rond from Darwin College for their kind invitation to give the lecture which formed the basis for this chapter, and Janet Gibson for her logistical help; Terry Burke, Ammon Corl, Iain Couzin, William Foster, Christos Ioannou, Rufus Johnstone, Rebecca Kilner, Theunis Piersma, and Richard Wrangham for discussion and information on the examples I discuss; Suzanne Alonso, Stuart Butchart, Kathelijne Koops, David Quinn, Dave Leech, Claire Spottiswoode, and Andrew Smiley for permission to reproduce illustrations; and the Natural Environment Research Council for funding my research.

---

[40] C. C. Ioannou, V. Guttal, and I. D. Couzin, 2012, Predatory fish select for coordinated collective motion in virtual prey, *Science* 337(6099), 1212–1215.

[41] I. D. Couzin, J. Krause, N. R. Franks, and S. A. Levin, 2005, Effective leadership and decision-making in animal groups on the move, *Nature* 433(7025), 513–516.

[42] I. D. Couzin, C. C. Ioannou, G. Demirel, T. Gross, C. J. Torney, A. Hartnett, *et al.*, 2011. Uninformed individuals promote democratic consensus in animal groups, *Science* 334(6062), 1578–1580.

Afterword

The Game Theory of Conflict
The Prisoners' Dilemma – An Unsympathetic Critique

THOMAS C. SCHELLING

Let me begin by telling a story. My wife and I visited Iran around 10 years ago; I went to the Sharif University of Technology in Tehran and they offered me an honorary degree. When I arrived, they suddenly discovered that their university, which had a Department of Economics, had not yet awarded its first PhD, and, according to their own rules, would not be able to offer an honorary degree if not already having awarded a PhD. The Economics Department despaired. Yet the University decided that, since they had shipped me all the way to Tehran from the USA, there must be a way around this. So they went to the Mathematics Department and said 'You know, this guy was awarded a Nobel Prize for game theory and you know what game theory is, don't you?' They said 'Yeah, we've heard of it', and so the University asked whether they could then perhaps award me an honorary PhD in mathematics. They happily obliged, meaning that I am actually an Honorary Doctor of Mathematics, despite not really being a mathematician. It is with those 'credentials' established that I turn to a mathematically derived puzzle with social implications.

In this brief reflection, which concludes a stimulating and wide-ranging volume on *Games*, I explore the most famous, most reproduced, most acknowledged, and most thought-about story in the history of game theory: the 'Prisoners' Dilemma'. It originates from unpublished notes made by Albert Tucker while he was spending some time at Stanford University. The dilemma is briefly this:

> Two men charged with a joint violation of law are held separately by
> the police. Each is told that, if one confesses and the other does not, the

former will be given a reward of one unit and the latter will be fined two units. If both confess, each will be fined one unit. At the same time, each has good reason to believe that if neither confesses both will go free.

That is all there is to it. Yet it is the most reprinted problem in the history of game theory, even though it was never published by its original author. Frankly, I do not think he spent much time pondering it. All he wanted to do was demonstrate that there are situations, very often but not always involving two parties, in which the best choice for each turns out to be not the best joint choice.

Tucker's Absent Assumptions

Now this is commonly understood as a dilemma in dealing with multi-party situations. For example, almost everybody involved in climate change debates realises how collectively agreeing to do something about climate change is by far and away the best thing to do compared with doing nothing, and yet, at the same time, it would be in the individual interest of any nation to sit still, forget about climate change, and let all other nations 'do the job' for them.

I am going to walk us through this dilemma in a way that its author, Albert Tucker, might not have expected. Consider, for example, the following dilemma: how can I borrow money from one of you and prove that I will pay it back? Society has, by and large, 'solved' this dilemma by developing institutions that permit people to borrow money, usually using something as collateral. The invention of the pawn shop is a good example: if I have a musical instrument, I can take it to a pawn shop, leave it there and borrow an amount of money up to a third of the value of, say, the clarinet, knowing that, if I do not pay back the loan with interest on time, the shop will sell the clarinet and keep the proceeds. Similarly, banks can sue for non-payment of loans or cash in whatever collateral you put up against the loan. If I try to order something by telephone or by mail and they want money in advance, do I trust them to ship me what I've paid for, or, if they ship it first, do they trust me to pay them for it eventually? These sorts of dilemmas no longer really exist for most of us because, as a society, we have found ways to resolve them.

Still, Albert Tucker's dilemma intrigues me, partly because he never went through the trouble of working out the analysis. Below, I will try to work out an analysis of the problem, as I see it.

The Prisoners' Dilemma suggests that if I, as one of two parties charged with a joint violation of the law, expect you to confess, then I too have to confess because if I do not, I will get fined doubly. If I expect you not to confess, then I gain by confessing because I collect a reward. And I think it goes without saying in the original version that confession is jointly confession for oneself and incrimination of the partner, so that all it takes is one confession to get us both declared guilty. And if he doesn't confess, but I confess, he is found guilty and is fined doubly. Therefore, it is in his interest, whatever he expects me to do, to confess. And it is in my interest, whatever I expect him to do, to confess. And yet, at the same time, each has good reason to believe that if neither confesses both will go free. That is why it is called a 'dilemma', even though I think dilemma is not quite the correct term insofar as a dilemma represents any choice between two equally offensive outcomes, and this is a case in which to each of us the two outcomes are not equally offensive, except as a result of our collective decisions.

Known Unknowns

Now here are a few things we do not know – but would want to know – about those involved in this dilemma. All we know about these two people is, in the original language, that they are 'two men' (that is, they are not husband and wife). They could be father and son, or brothers, or members of a fraternity, or of the same church, or even of a well-integrated but isolated immigrant group. Their connection – or lack thereof – is relevant insofar as it might allow us to predict whether they are, or are not, likely to trust each other not to confess.

The easiest case to consider is that in which my brother and I are jointly arrested and we expect if neither confesses we'll go free, but if one confesses the other had better confess also. However, we do not know whether these people knew each other because part of what Tucker said was 'It also follows that if the two could reach a binding agreement they would readily agree, obviously agree not to confess but they being separated that result is not available.' If they weren't separated they could always agree on something and shake hands, they wouldn't necessarily then trust each other unless the two are brothers or father and son or something of the sort. On the other hand, I think it is very likely that, if two people knew each other well enough to be arrested together and charged with having pulled off a successful heist of some sort, they may be intimate enough with each other that they won't need an enforceable agreement; they might just agree 'Let's not confess, neither of us, then we'll go free.'

Perhaps the men act as Boy Scouts might, the oath being 'On my honour I will do my best to do my duty to God and my country and to obey the Scout Law.' This 'Scouts' Law' consists of twelve commandments, the first of which is that a Scout is trustworthy, the second that a Scout is loyal. If the two men are trustworthy and loyal, they wouldn't have to say 'Scout's honour', but would instead simply recognise that neither would snitch on the other just to collect the double reward.

It is not immediately obvious that merely by virtue of their being separated they are not in a position to enter into a binding agreement. This is because there are many ways to bind an agreement, and thus two people could easily arrive at an agreement that neither intends to violate even without consulting each other first.

The question remains whether each can trust the police who promised to provide a reward to the one who confesses if the other does not. What guarantee is there that they will?

Moreover, are the men given access to lawyers? If they are, their lawyers could presumably consult each other and act as brokers such that they find ways by which to commit each man to cooperation, or silence. Perhaps the lawyers might be able to get the police to commit their proposal and pay-off structure to paper to be signed off by someone in authority.

If they manage to obtain a signed document, the question then arises as to whether their guilt (and, if so, penalty) will be determined by a judge or by a jury. If we assume that a judge will have the authority to decide this case, confessions have to go before the judge. But if the men (or their lawyers) take this signed document to a judge they could legitimately argue that the prosecutor is guilty of two heinous crimes: bribery and extortion. Thus, I doubt the prosecutor would be willing to put his name to the proposal and pay-off structure. It may even be that he hasn't the authority to double the fine on whomever refuses to confess (assuming the other does confess), making the proposal a mere empty threat.

Finally, if the confession of one (and silence of the other) sees one being sent home with extra cash in his pocket, and the other off to prison for twice the ordinary term (or out of pocket by twice the monetary fine), it might be that – even if the men have no prior relationship – they might live in the same neighbourhood, attend the same church, or have children attending the same school. Either way, people will find out that one snitched on the other. In this case, there are powerful, and predictable, social mechanisms that may ensure cooperation even among people who otherwise have no relationship, or punishment following the incrimination of one by the other.

Conclusion

Thus, context is everything, and without the context it is hard to know how people might 'solve' even relatively simple games. The questions presented in this brief reflection are not all new innovations in game theory, but they illustrate a crucial point: that 'games' do not 'speak for themselves', but rather must be located in the social environment in which they arise if they are to have meaning or offer insight. I doubt that Albert Tucker ever imagined the difficulties his Prisoners' Dilemma might produce for future generations; indeed, I am fairly certain he gave it little or no more thought. I hope, however, that readers of this volume will continue to do so, because it is only by thinking through the human contexts of such games that we will be able to deliver future solutions to important cooperation problems.

Index